Geographies of Commoᵈ⸍ Chains

At the beginning of the twenty-first century there is increasing concern about the intricate relationships between people, places and commodities. Individuals, consumer groups, nation states and supra-national bodies increasingly are interrogating the ethics of particular production and consumption relations. Flowing from and bound up with these political concerns is the growing interest in the mutual dependence of sites of production, distribution, retailing, design, advertising, marketing and final consumption.

Geographies of Commodity Chains brings together a wide variety of contributions to examine critically the spaces through which consumers are connected to producers. Case studies are drawn from a broad range of commodities including fruit, clothing and furniture, and span geographical areas from Europe, to America and Africa. Not only do these examples seek to transcend older understandings of production and consumption, but also they explicitly tap into wider public debate about the meanings, origins and biographies of commodities.

This book takes a geographical approach to the analysis of links between producers and consumers, focusing upon the ways in which these ties increasingly are stretched across spaces and places. Critical engagements with the ways in which these spaces and places affect the economies, cultures and politics of the connections between producers and consumers are threaded through each section.

Alex Hughes is Lecturer in Geography at the University of Newcastle, UK and **Suzanne Reimer** is Lecturer in Geography at the University of Hull, UK.

Routledge Studies in Human Geography

This series provides a forum for innovative, vibrant, and critical debate within Human Geography. Titles will reflect the wealth of research which is taking place in this diverse and ever-expanding field.

Contributions will be drawn from the main sub-disciplines and from innovative areas of work which have no particular sub-disciplinary allegiances.

Geographies of Commodity Chains

Edited by
Alex Hughes and
Suzanne Reimer

 Routledge
Taylor & Francis Group

LONDON AND NEW YORK

First published 2004
by Routledge
2 Park Square, Milton Park, Abingdon, Oxfordshire OX14 4RN

Simultaneously published in the USA and Canada
by Routledge
711 Third Avenue, New York, NY 10017

Routledge is an imprint of the Taylor & Francis Group

First issued in paperback 2011

Typeset in Galliard by
Newgen Imaging Systems (P) Ltd, Chennai, India

British Library Cataloguing in Publication Data
A catalogue record for this book is available
from the British Library

Library of Congress Cataloging in Publication Data
Geographies of commodity chains/[edited by] Alex Hughes and
Suzanne Reimer.
p. cm. – (Routledge studies in human geography; 10)
Includes bibliographical references and index.
1. Commercial products. 2. Commercial geography.
3. Consumption (Economics) – Moral and ethical aspects.
4. Produce trade – Environmental aspects. 5. Physical
distribution of goods. 6. Production (Economic theory)
I. Hughes, Alex (Alex Louise) II. Reimer, Suzanne. III. Series.

HF1040.7.G46 2004
306.3'4–dc22   2003024890

ISBN 978-0-415-33910-0 (hbk)
ISBN 978-0-415-51403-3 (pbk)

Contents

Illustrations

Plates

Figures

Tables

Contributors

Hazel R. Barrett is Reader in Development Geography, School of Science and the Environment, Coventry University, UK.

Angela W. Browne is Principal Lecturer in the School of Science and the Environment, Coventry University, UK.

Ian Cook is Lecturer in the School of Geography, Earth & Environmental Sciences, University of Birmingham, UK.

Justine Coulson works in London for the Save the Children Foundation.

Philip Crang is Reader in Human Geography at Royal Holloway, University of London, UK.

Louise Crewe is Professor of Geography in the School of Geography, University of Nottingham, UK.

Niels Fold is Associate Professor in the Institute of Geography, University of Copenhagen, Denmark.

Julie Guthman is Assistant Professor in the Department of Community Studies, University of California, Santa Cruz, USA.

Alex Hughes is Lecturer in Geography at the University of Newcastle, UK.

Brian W. Ilbery is Professor of Human Geography in the School of Science and the Environment, Coventry University, UK.

Deborah Leslie is Associate Professor in the Department of Geography, University of Toronto, Canada.

Charles Mather is Senior Lecturer in the School of Geography, Archaeology and Environmental Studies, University of the Witwatersrand, Johannesburg, South Africa.

Mara Miele is a researcher in the Dipartimento di Agronomia e Gestione dell'Agroecosistema, Università di Pisa, Italia.

Carol Morris is a Senior Research Fellow in the Countryside and Community Research Unit, University of Gloucester, Cheltenham, UK.

Jonathan Murdoch is Professor in Environmental Planning, Department of City and Regional Planning, Cardiff University, Wales, UK.

Parvati Raghuram is Senior Lecturer in the Department of International Studies, Nottingham Trent University, Nottingham, UK.

Suzanne Reimer is Lecturer in Geography at the University of Hull, UK.

Petrina Rowcroft is a postgraduate student in the School of Geography, Archaeology and Environmental Studies, University of the Witwatersrand, Johannesburg, South Africa.

Mark Thorpe is Head of Qualitative Research at SPA, a London-based independent market research company.

Michael J. Watts is Professor of Geography and Director of the Institute of International Studies, University of California, Berkeley, USA.

Craig Young is Senior Lecturer in Geography, Department of Environmental and Geographical Sciences, Manchester Metropolitan University, Manchester, UK.

Acknowledgements

The author and publishers would like to thank the following for granting permission to reproduce images in this work:

Sainsbury's Supermarkets Ltd for Plate 4.1
Nestlé UK for Plate 9.1
Geest plc for Plate 9.2
Twentieth Century Fox/Kobla Collection for Plate 9.3
The Guardian Media Group for Plate 10.1 and Figure 10.1.

Every effort has been made to contact copyright holders for their permission to reprint material in this book. The publishers would be grateful to hear from any copyright holder who is not here acknowledged and will undertake to rectify any errors or omissions in future editions of this book.

Introduction

Alex Hughes and Suzanne Reimer

Discussions of the connections along commodity chains are becoming ever more prominent. Particularly within nations of the West, media and popular accounts of commodities increasingly have focused on the complex processes through which goods are reworked as they move through stages of design, production, advertising, marketing, retailing and final consumption. The pursuit of commodity stories has been prompted by a wide range of factors. These include the theoretical preoccupations of academics interested in nuancing previous understandings of relationships between production and consumption; the politically motivated concerns of activists seeking to illuminate disparities between low-waged producers and affluent consumers; and consumers' growing interests in the origins and provenance of the goods they buy. However motivated, an increased concern with the dynamics of the production/consumption nexus has stimulated a greater awareness and questioning of the intricate relationships between people, places and commodities in the early twenty-first century.

Geographies of Commodity Chains explicitly draws upon the conceptual framework of the commodity chain in order to critically examine the spaces through which consumers are connected to producers. A wide-ranging set of case studies traces geographies of commodities constructed via trading links, organizational networks and cultural and economic knowledges. Our contributors draw upon case study research covering a broad variety of commodities (fruit, vegetables, meat, cocoa, nuts, fast food, clothing, flowers and furniture) and different geographical areas (including the national contexts of the United Kingdom, Kenya, The Gambia, the United States, Ghana, Italy, South Africa, Ecuador and Canada). Although all of the authors centrally are concerned with links between moments of production, circulation and consumption of commodities, they adopt varying perspectives towards the understanding of producer–consumer relationships. As a backdrop to the chapters which follow, our discussion below first sketches out the broad contours of differing approaches. The remainder of the introduction reflects upon some of the key political and geographical debates which have emerged within the (growing) literature on commodity chains and provides an overview of each of the chapters. A final section identifies several methodological issues for future research.

Theoretical perspectives, approaches and metaphors

Global commodity chains and systems of provision

At a general level, the notion of the commodity chain is one of the most pervasive metaphors for thinking about the links between the production, distribution and consumption of goods (Watts, 1999). A central focus is upon the progressive movement of a commodity through the sequential phases of production, distribution and consumption. In the main, studies which have adopted the metaphor of the chain can be divided into two sets of literatures (see also Leslie and Reimer, 1999). With its conceptual roots in world systems theory, the linear connections of the global commodity chain (GCC) approach foreground commodity production in 'peripheral' regions of the world-economy for retail and consumption in the 'core' (Leslie and Reimer, 1999; see in particular work by Appelbaum and Gereffi, 1994; Gereffi, 1994a,b; Gereffi *et al.*, 1994; Hopkins and Wallerstein, 1986). As Gereffi *et al.* (1994, 2–3) argue, 'enterprises and states in the core...gain a competitive edge through innovations that transfer competitive pressures to peripheral areas of the world-economy'. Such competitive pressures often are manifest in the form of low-wage labour and economic insecurity, primarily shaped by short-termist programmes of economic development centred upon export-led growth, debt-alleviation and structural adjustment policies (Gereffi, 1994a,b; McMichael, 1998).

There is a noticeable overlap between the GCC approach and political-economic accounts of agriculture (for the latter, see Friedmann, 1993; Friedmann and McMichael, 1989; McMichael, 1998; Raynolds, 1997; Watts, 1992; Watts and Goodman, 1997). The influence of political economy approaches is perhaps most visible in the use of the term filière (particularly when it appears interchangeably with commodity chain). The notion of a filière has its origins in French industrial economists' investigations of price formation in the journey of a commodity from raw material to final product, through various stages of physical transformation (processing, manufacturing), transport, storage and so on (Raikes *et al.*, 2000). Like accounts of the global commodity chain, discussions of filières signify a highly economic conceptualization not only of the nature (i.e. transformations of value) but also the form (linear and tightly strung together) of relationships between different sites.

Perhaps one of the most useful aspects of the GCC approach is provided by Gereffi's (1994a,b) characterization of the 'modes of organisation' that govern the commodity chain. Having distinguished between producer-driven and buyer-driven chains, Gereffi draws attention to the importance of the buyer-driven chain in driving trade in consumer goods such as garments, footwear, toys, consumer electronics, housewares and hand-crafted items. Emphasis is placed upon the power of commercial and industrial capital in allocating economic resources and wealth within – and therefore defining the shape of – global commodity chains.

The GCC approach has, however, received some criticism. Just as Cook *et al.* (1996) suggest that macro-scale political economies of agriculture emphasize large-scale food systems at the expense of the cultural richness of localities, Leslie and Reimer (1999) point to the reductionist nature of the systemic links in the production–consumption chain advocated by the GCC literature. Furthermore, the GCC approach primarily focuses upon adverse economic effects felt by sites of production in the periphery of the world-economy and tends to treat retail and consumption sites as simple starting points from which to embark on a more worthy examination of exploitation in the productive sphere (Leslie and Reimer, 1999).

Some of the shortcomings of the GCC approach are partially redressed in a second set of literatures focusing upon 'systems of provision' (Fine and Leopold, 1993). Systems of provision accounts have been seen to offer a 'more balanced treatment of the relationship between production and consumption' (Leslie and Reimer, 1999, 11) than are provided by the GCC tradition. Studies of systems of provision promote a more dialectical understanding of the producer–consumer relation, recognize the importance of the cultural meanings attached to commodities and appreciate that the producer–consumer dynamic can be different for contrasting industries. However the approach still tends to prioritize the significance of the sphere of production, while underscoring more systemic connections between production and consumption (Leslie and Reimer, 1999).

GCC and systems of provision approaches importantly highlight links between production, distribution and consumption. They provide insights into economic effects at sites of production and – in Gereffi's (1994a,b) analysis – highlight the buyer-driven mode of governance as a significant causal mechanism. However analytical constraints derive from the often unidirectional linearity of approaches as well as from a culturally and geographically impoverished nature of system-based methodologies. GCC and systems of provision approaches may overlook the complex practices through which production and consumption are linked and often fail to consider precisely how buyers may control and condition the economic fortunes of 'the periphery'.

Commodity circuits

A contrasting way of understanding interconnected production, circulation and consumption processes is afforded by the literature on commodities and circuits of culture (Cook and Crang, 1996; Cook *et al.*, 1996; Jackson, 1999; Jackson and Taylor, 1996; May, 1996). Circuits of culture approaches treat the movement of commodities through phases of production, distribution and consumption as a non-linear circuit, rather than a linear chain (Leslie and Reimer, 1999). Rather than focusing upon beginning and end points in a chain, attention is directed towards the culturally inflected dynamics of relationships between moments of production, circulation and consumption. Embedded in a much broader literature on commodity cultures, the aim is to arrive at more contextual understandings of meanings attached to goods in different times, places and phases of commodity circulation. Circuits of culture analyses are underpinned by developments in economic anthropology, studies of material culture and critical ethnography (Appadurai, 1986; Bell and Valentine, 1997; Carrier, 1990; Howes, 1996; Kopytoff, 1986; Miller, 1997, 1998; Weiss, 1996). In contrast to approaches to the commodity chain that aim always to 'unveil' economic processes at sites of production, both Crang (1996) and Jackson (1999) suggest that it is more fruitful to consider the complex ways in which goods may be displaced from one site to another and the ways in which various cultural and geographical knowledges inform and are informed by processes of displacement. Rather than prioritizing 'realities' at the site of production and searching for causal mechanisms responsible for these realities, work on commodity cultures aims to thicken descriptions of the meanings attached to goods through a cultural analysis of different phases of commodity circulation, including consumption. Studies of circuits of culture specifically examine how meanings attached to goods are transformed within and across different phases of commodity circulation – for example between producers and consumers of advertising (Jackson and Taylor, 1996).

Commodity networks

A further means of rethinking relationships between production and consumption is provided by the concept of the *network*. While this metaphor has been widely adopted in the social sciences, it has had more limited application to studies of commodities. However it offers to resolve some of the contrasting difficulties associated with both commodity chain and circuits of culture approaches to understanding the producer–consumer relation. In broad terms, the idea of the network helps to conceptualize the complex and multi-stranded ways in which different types of nodes (people, firms, states, organizations etc.) are connected (Dicken *et al.*, 2001; Thrift and Olds, 1996). While Thrift and Olds (1996) explain the use of the network as a representation of the organization of social and cultural ties in economic linkages, they suggest that at the most general level of analysis it simply captures the pattern of 'webs of interdependence' existing between different sets of actors in the economy (Powell and Smith-Doerr, 1994). Inspired by concepts of the nervous system in biology and electrical networks in physics and engineering, Thrift and Olds (1996, 322) argue that 'the topological presupposition of the network is now in common usage in the social sciences as the emblem of an ambition to produce flatter, less hierarchical theories of the economy'.

Applying the metaphor of the network most simply to the morphology of relationships between production, distribution and consumption, it becomes possible to move away from the unidirectional linearity of links promoted by commodity chain approaches. While still recognizing the existence of sets of actors (or nodes) whose work it is to shape the circulation of a particular commodity, connections between actors are seen as complex webs of interdependence rather than fixed, vertical and unidirectional relationships. Such webs not only connect firms through vertical commodity exchange relationships, but also bind together additional agents through the multi-directional flows of information and materials that variously support these exchange relationships. As such, the privileging of one sphere of commodity circulation over another is ideally avoided, as the network metaphor is extended to include, for example, sites of design, research and development, non-governmental organizations and consumer groups (Hughes, 2000).

The analytical capacity of the network can be extended through the development more formally of an actor-network approach, which has its origins in the new sociology of scientific knowledge. Two of the central tenets of actor-network theory (ANT) are worth discussing here, in so far as they inform key developments in ideas about commodity networks. The first is that agency in a network is driven by both humans and non-humans. In other words, material objects have a transformative capacity within the networks of which they are a part (Latour, 1991, 1993; Murdoch, 1997; Pile and Thrift, 1995). One example of research which draws upon ANT specifically to develop the concept of the commodity network is provided by Busch and Juska's (1994, 1997) study of rapeseed production. The authors argue that the extension of production networks has, at least in part, resulted from the modification of relationships between people and plants through the development of rapeseed technoscience.

A second tenet important to discussions in this book is actor-network theory's conceptualization of the spatial morphology of networks. Put simply, ANT dictates that networks are always localized, working in real places and at specific times. As such, networks can only be made knowable by accounts of their workings 'on the ground'; and networks can only be considered as globalized in terms of their physical extension across space in practice (Murdoch, 1997). From the perspective of ANT, the globalization of a commodity network is considered primarily in terms of the extended spatial reach of associations between its key agents. This geographical slant on the morphology of commodity networks also is taken up by Busch and Juska (1997) in their study of the globalization of rapeseed production networks. Busch and Juska (1997) suggest that contemporary technologies permit new forms of 'action at a distance' that pave the way for extending the physical distances over which production networks can be practised. In tracing the connections of fair trade coffee networks, Whatmore and Thorne (1997) similarly consider 'network lengthening'. Here, the concept of network lengthening fruitfully problematizes binary notions of core and periphery used in both the macro-scale political economies of food and the GCC approach. Rather, Whatmore and Thorne (1997, 291) suggest 'the power associated with global reach has to be understood as a social composite of the actions and competences of many actants'.

Network-inspired analyses recognize that relations between producers, distributors and consumers are the product of complex flows between hosts of interconnected actors that have become enrolled in the network. Network-informed accounts thus might at least partially resolve concerns about the failure of a circuits of culture approach both to locate institutional power and to stress the forces responsible for exploitation at sites of production (see Leslie and Reimer, 1999). Rather than explaining competitive pressures at sites of production in terms of unidirectional causal mechanisms suggested by the GCC approach, explanations are traced through multi-stranded connections forged between interrelated actors.

Political and geographical debates

As the preceding discussion implies, different approaches to the study of commodity chains exhibit contrasting political stances. For Marxist-inspired accounts of global commodity chains and filières, materialist understandings of geographies at the site of production are seen to be crucial to the politics of consumption. Authors such as Harvey (1990) and Hartwick (1998, 2000) have exhorted academics and activists to become involved in exposing previously 'hidden' aspects of commodity production. Other commentators have questioned the politics of such an approach – Jackson (1999, 98), for example, is disquieted by 'the implication that academics have a uniquely critical insight into the social relations and conditions of production that escape the notice of "ordinary consumers"'. The development of commodity circuit approaches thus can be seen – at least in part – as a response to some of the political limitations of understanding commodity fetishism as a relatively straightforward process of 'veiling' (see also Cook *et al.*, Chapter 9, this volume). Notions of 'unveiling' and 'unmasking' are seen to render consumer knowledges and imaginaries as uninformed and inconsequential (Jackson, 2002b, 293).

Seeking to remedy information 'deficits' also might be seen as a problematic strategy from the perspective of other actors across the chain or network. A UK journalist's recent investigation of the footwear industry, for example, 'revealed' that the production origins of brogues stamped as 'Made in Britain' could be traced to sub-assembly processes conducted in a town on the south-east coast of India (Abrams, 2002). Part of Abram's narrative involved informing a woman worker in Chennai that the shoes to which she attaches eyelet facings are sold in the United Kingdom for the equivalent of the wage she earns in three months. Here, the extension of information across the commodity chain (in this case, *towards* producers and production locations) seems at best a superficial and at worst a disruptive tactic. This is not necessarily to decry Abrams's political motivations in presenting shoe stories to British readers of *The Guardian*. Rather, it is to argue that calls to 'unveil' commodity fetishism must be much more attentive to the practical and moral implications that such tactics might have across a range of geographical locations and spatial scales (see also Castree, 2001).

Circuit-inspired accounts themselves, however, also have been called into question. Leslie and Reimer (1999, 407), for example, have argued that the notion of an endless circuit

> may involve the loss of an important political stance: the foregrounding of exploitation. It is for this reason that we are hesitant to abandon the concept of the chain altogether...[I]f the aim of commodity chain analysis is no longer to determine what forces are driving the chain, we are left with a question as to why chains should be reconstructed at all.

In response, Jackson (2002a, 15) recently has argued 'demonstrating complexity is scarcely a worthwhile end in itself'. He has reiterated an interest in explicitly identifying 'sets of tensions and anxieties' (after McRobbie, 1997) across the commodity network – precisely as a means of locating points of political connection and debate in production, retail and consumption spaces. Further, Jackson suggests that analysts must carefully consider '*all* the points of connection within a network or throughout an circuit rather than [...] successive points down a linear commodity chain' (2002a, 15, our emphasis).

Debate about the possibilities for (re)connecting producers and consumers generally has tended to rely upon case study examples in which the two sets of actors are geographically distant. Within athletic footwear production, the fashion and clothing industries and the fruit and vegetable trade, commodities frequently travel vast distances across the globe. Perhaps as a result, much recent theoretical discussion of commodities has dovetailed with debates about globalization (for a helpful reflection upon this situation, see Jackson, 2002b). For our purposes here, however, we suggest that future geographical debates about commodity chains appropriately might be advanced by attention to aspects of production–consumption dynamics which are not *necessarily* stretched across global spaces. For example, rather than focussing directly upon producer–consumer linkages, Mutersbaugh's recent discussion of organic coffee certification explicitly recasts a conceptual net 'to capture producer–certifier–labeller links' (2002, 1166). Albeit emphasizing constraints placed upon producer organizations by Northern certification norms (predominantly deriving from the European Union and the International Organisation for Standardisation (ISO)), Mutersbaugh's (2002) account also importantly illuminates the dynamics of management and inspection practices at the level of individual producers and producer organizations. Drawing attention to the complexities of sites other than production and consumption 'ends' of the chain also may be important if we are to avoid reductionist distinctions between Northern consumers and Southern producers (see Raghuram, Chapter 6, this volume).

Advancing debates: the structure of the book

Geographies of Commodity Chains brings together a collection of chapters which, in various ways, both challenge and take forward existing theories, approaches and metaphors used to capture relations between production and consumption. The book is divided into four interconnected parts: (1) Commodity chains, networks and filières; (2) Commodity chains and cultural connections; (3) Commodities, representations and the politics of the producer–consumer relation and (4) Ethical commodity chains and the politics of consumption.

Commodity chains, networks and filières

The central theme of the first part is the organization of vertical links in the commodity chain. In evaluating the governance of particular commodity chains, authors adopt a predominantly political-economic approach, although each chapter utilizes a slightly different metaphor to represent commodity linkages. Barrett *et al.* problematize existing approaches to the study of commodity link-ages in the context of research on the international trade in horticultural produce. Their study considers trading connections between UK supermarket chains and producers of fresh fruit and vegetables in Kenya and The Gambia. Research by Barrett *et al.* incorporates links between key trading sites: retail and consumption, import, export and production. The authors argue that knowledge, information and the degree of embeddedness of actors in trading links together determine the economic success of producers. It is suggested that Kenyan producers, given their longer history of involvement in the horticultural export business, capitalize more effectively on knowledge networks than their counterparts in The Gambia. Critically, Barrett *et al.* argue that none of the three main approaches to the geography of commodities – commodity chains, commodity circuits, commodity networks – is on its own capable of capturing these knowledge-intensive trading links. While they suggest that the network metaphor is the most appropriate con-ceptual tool for theorizing connections between the different trading sites, they propose that it should be combined with an appreciation of commodity circuits to envision flows of information and knowledge between the sites. They therefore forward a model of a 'cellular network', combining network and circuitry imaginaries, as a means of theorizing global commodity linkages.

Watts also attends to the organizational dynamics of commodity linkages, this time in the context of the US chicken and pork industries. In assessing the degree to which the organization of the hog sector is replicating that of the broiler, and thereby displaying commodity convergence, he contributes to a long-running debate concerning the difference particular commodities themselves make to the organizational dynamics of the chain (see, e.g. Fine and Leopold, 1993; Friedland *et al.*, 1981). Drawing upon accounts of captive animals as well as notions of confinement via enclosure, Watts critically examines the impacts of mechanization on both the broiler and hog filières. While the hog industry appears to be following the organizational trajectory of the broiler sector, Watts also illuminates the historico-geographical and biological specificities of each commodity chain.

In a related vein, Fold's examination of cocoa bean and shea nut production in Ghana also explores questions of similarity and difference across two agricul-tural commodity chains. Crucially, commodity chain dynamics in Fold's case study context have been sharply inflected by restructuring processes initiated by the demands of Structural Adjustment Programmes. The story of export com-modities such as cocoa beans and shea nuts thus emphasizes the importance of regulatory contexts in constructing and reconstructing commodity chains. Further, Fold notes that the liberalization of the Ghanaian cocoa chain has led to a decline in locally managed quality control systems – at precisely the same time that Northern chocolate manufacturers and consumers increasingly have begun to demand 'traceability'. Drawing attention to a theme which threads its way through a number of the chapters to follow, Folds considers that the future imposition of detailed quality assurance systems by agencies and manufacturers in the North might will have a negative effect upon African producers.

Commodity chains and cultural connections

The thematic pivot of the second part of the book dovetails with a disciplinary debate that has been shaping economic geography for close to a decade. The so-called 'cultural turn' has placed image-creation, identity-construction and performance much more centrally on the research agenda of the sub-discipline. While there has been intense debate over the utility of taking cultural dimensions of the economy more seriously (see, e.g. Barnes, 1995; Sayer, 1994), constructive commentaries and applied research have suggested the importance of the ways in which 'the cultural' and 'the economic' are intertwined and mutually constitutive of one another (Barnes, 1995; Crang, 1997; McDowell, 1998; Schoenberger, 1997 and from a more interdisciplinary perspective, du Gay and Pryke, 2002). However, Jackson (2002a, 4) has most recently suggested that 'calls to transcend the "great divide" between the cultural and the economic have significantly outnumbered empirically grounded studies that demonstrate the difference that such a move would make in practice'.

Each of the chapters in Part III takes up this challenge by exploring the centrality of culture – in the form of meanings, aesthetics, creativity and imagination (Jackson, 2002a) – to the competitive dynamics and organization of commodity chains and networks. Morris and Young focus on the production and consumption of knowledge used in the United Kingdom food industry's quality assurance schemes. These schemes have come to form a central part of the private interest re-regulation of the food system, and are pivotal to the maintenance of competition in the food sector (Marsden *et al.*, 2000). However notions of food and production quality are also socially constructed. Adopting the conceptual framework developed by Cook and Crang (1996) regarding geographical knowledges of commodities, Morris and Young evaluate representations of quality within assurance schemes. Their approach reveals how at first sight, schemes appear to be reconnecting producers and consumers by providing more information on the production process and product origins for the consuming public. However Morris and Young also suggest that quality assurance schemes are in fact prone to obscuring processes of production and distribution for consumers, which may serve to fuel consumers' ambivalence about the food system. Such a conclusion problematizes the simple 'unveiling' of production 'realities' suggested by Marxian approaches to the commodity chain (Hartwick, 1998; Harvey, 1990) and foregrounds instead the merits of a commodity circuits approach to the politics of agro-food supply chains.

Retaining a focus on the significance of 'the cultural', but taking a quite different theoretical approach, Murdoch and Miele examine geographies of food consumption and culinary networks. Through a new intervention in existing theories of commodity linkages, they argue for a conceptual approach that combines ANT with conventions theory. While the former is used to capture the web of heterogeneous associations between human and non-human actants in a provisioning system, the latter attends to the rationalities and cultural norms that underpin commodity networks. The authors provide a comparative analysis of two contrasting culinary networks: fast food, epitomized by McDonalds, and the Slow Food movement. The Slow Food movement is recognized as resisting the culinary cultures of McDonalds, with its conventions of market performance, industrial efficiency, 'reknown' and civic equality alongside highly mediated and distanciated relationships between producers and consumers. Slow Food seeks to reconnect consumers much more directly with locally based systems of food provision. Murdoch and Miele suggest that the growth of each culinary network is driven by conventions emanating from their cultural and geographical origins (the suburban United States in the case of McDonalds and the locally derived food traditions of Italy in the case of the Slow Food movement). At the same time, however both networks coexist in the global economy.

While Murdoch and Miele have sought to advance debate by blending elements of actor-network and conventions theory, Raghuram's critique of existing commodity chain approaches shifts direction to consider the role of individual actors in initiating networks of production. Raghuram places the agency of South Asian women entrepreneurs at the centre of her analysis. While rejecting the global commodity chain approach pioneered by the work of Gereffi (1994a,b) on the basis of its (over)emphasis on the activities of large global corporations, Raghuram also is sceptical of the sole focus on product biographies often found in more culturally inflected accounts. Against the casting of 'Third World Women' workers as passive figures in commodity chain stories, Raghuram argues for an approach which is attentive to the personal biographies of actors. Through an investigation of the ways in which the recognition of a market niche as a (British) consumer led one South Asian woman to become a producer of garments, Raghuram challenges the analytical separation of production and consumption in global commodity chain approaches. She argues that attending to the personal biographies and cultural identities of key actors in the clothing commodity chain will enable analysts to 'find a place for South Asian women at the nexus of production and consumption tales'.

Commodities, representations and the politics of the producer–consumer relation

The third part of the book maintains a focus on commodity cultures whilst beginning to explore more explicitly the politics of producer–consumer linkages. Each chapter in this part advances a commodity circuits approach, focusing in different ways on the geographical knowledges circulating between interconnected actors in spheres of production, distribution and consumption (Cook and Crang, 1996; Crang, 1996). Representations of commodities in advertisements, product promotions and campaign images form the critical focus of inquiry. Coulson explores geographical knowledges mobilized in advertisements for Ecuadorian cut flowers. She critically analyses the representation of people, landscape and product in advertisements appearing in the cut-flower trade press during the mid-1990s. Coulson stresses the importance of representations of bodies within advertisements and reflects upon the inscription of these represented bodies with particular ethnic and gendered identities.

Mather and Rowcroft's chapter represents a significant political intervention in the literature on commodity geographies. The authors suggest that considerations of geographical knowledges of commodities often have focused solely on advertising imagery, despite Crang's (1996, 54) assertion that knowledges of commodities 'are constructed through rather more discursive fields'. In the spirit of redressing this imbalance in the literature, Mather and Rowcroft investigate the contested meaning of Outspan oranges and the embeddedness of this struggle in both South African domestic politics as well as international relations. While Outspan's promotional campaigns overseas during the 1960s and 1970s avoided any geographical references to product biographies and origins, the anti-apartheid movement in Europe produced campaigns directed at consumers which linked the branded Outspan product directly with apartheid and the social relations of production in South Africa. In the post-apartheid era, the authors argue that Outspan's use of pan-African imagery, focusing on idealistic representations of the natural landscape, is a deliberate strategy to avoid direct association with the country's specific political past.

Cook *et al.* also address the politics of the producer–consumer relation, applying and developing earlier ideas on geographical knowledges and commodity fetishism (Cook and Crang, 1996). Their empirical focus is upon the commodity fetishism of exotic fruits sold in the United Kingdom. They begin by discussing some of the fetishistic knowledges of the tropics used to advertise and market fruits in the UK trade press. However rather than suggesting that we should cast aside the 'veil' of this imagery in order to reveal the 'true' social relations of production, Cook *et al.* argue that we can instead attend to the ruptures *within* the fetish. The authors identify ruptures within the ways in which promotional knowledges used to sell tropical fruit are framed in part by European colonial discourses circulating through the fifteenth to the nineteenth centuries. Further, they suggest that there is much to be gained politically by engaging with the fetish itself rather than simply sweeping it aside. Drawing on discussions of Carmen Miranda in the 1940s, Cook *et al.* propose that ruptures in the commodity fetish can in turn be recontextualized in ways that might be politically progressive.

Ethical commodity chains and the politics of consumption

The final part of the book continues to tackle the politics of the producer–consumer relation through a specific focus upon aspects of so-called 'ethical' trade. Crewe's chapter reflects upon contemporary transformations within UK high street fashion retailing and in particular upon the recent abandonment of domestic sourcing in favour of low cost overseas suppliers. Shifts in the geographies of fashion, Crewe argues, have had important implications for the politics of consumption. As Northern consumers' geographical imaginations extend to take into view sites across the commodity chain, the practices of clothing retailers increasingly have been called into question. However ethical trading initiatives seeking to regulate labour conditions in the globalized fashion industry are seen to be weakened in part because of the complexity of the intricate webs in which production, subcontracting, retailing and consumption sites are bound up.

Hughes' examination of ethical trade in the cut flower industry also interrogates emergent forms of regulation across the commodity chain. The ethical trading movement has come to depend upon the detailed auditing of production, in which codes of practice in the cut flower industry have been developed primarily to enhance retailer profitability (often seeking to counter negative media exposure of pesticide use and labour exploitation) and as such are highly abstracted from the lived experiences of workers. Hughes' theorization of the dynamics of ethical trading is indebted to recent discussions of virtualism: she suggests that the cut-flower commodity chain crucially has been shaped by the actions of UK supermarkets as 'virtual consumers' as well as by the disembodied knowledges which underpin codes of practice in the industry.

Drawing upon research conducted in California, Guthman considers the politics of consumption which has accompanied the growth of 'organic' food provision. She suggests that whilst at one level the labelling of 'organic' products might be seen to provide a means of defetishizing commodities, the regulation of organic food in fact has resulted in a conflation of public choice and civic protest with consumer 'choice' and profit-making. In Guthman's view, the production- and input-oriented rules of the certification process act to *re*-mystify commodity meanings, given that certification and labelling ultimately obscure many aspects of commodity chain dynamics. Guthman is particularly concerned about the ways in which the 'branding' of food as organic fetishizes processes of social change – that is, the act of purchasing organic food itself has come to stand for effecting broader economic and societal transformation.

Discussion of ethical aspects of consumption continues in Chapter 13 in which Reimer and Leslie investigate the decision-making processes of home furnishings consumers in the United Kingdom and Canada. Returning to debates about the distinctiveness of individual commodity chain dynamics, Reimer and Leslie suggest that levels of consumer concern about the provenance of furniture are significantly shaped by the commodity's particular relationship to the body as well as its inherent qualities of durability and longevity. Further, the authors argue that although large global retailers such as Ikea recently have sought to engage with (potential) consumer concerns about origins (particularly in the case of wooden furniture), the introduction of ethical codes in the furniture industry primarily functions as a marketing strategy rather than as a means of transforming production processes and conditions. Theoretical critique in Reimer and Leslie's chapter is directed towards both Marxian and actor-network approaches to the commodity chain. In contrast to Guthman, Reimer and Leslie are less convinced by explanations which rely upon characterizations of mystification and masking of commodity origins, for they reject the idea that there is a singular, 'true' story to be told about associations between producers and consumers. At the same time, they are concerned that actor-network approaches often underplay the differential power of agents in a network. The challenge for future theoretical and political interventions in debates about ethical trade is simultaneously to accord sufficient weight to the influence of different actors/actants *as well as* network rules and conventions (see also Murdoch and Miele, Chapter 5, this volume).

Final thoughts

Each of the contributors to this book has been attracted to conceptual frameworks that take as their starting point the interconnections between sites of production, circulation and consumption. As the collection demonstrates, interventions in understandings of the commodity chain help to move beyond relatively compartmentalized understandings of production versus consumption, economy versus culture and the material versus the symbolic (Jackson, 2002b, 295). Our final thoughts briefly reflect upon the adoption of a commodity chain approach as a methodological strategy – something which we feel can be inspiring as well as enormously challenging!

Many of our authors have sought to extend fixed and linear conceptualizations of the commodity chain. Although analysts increasingly are reluctant to identify beginnings and ends of single-stranded chains, theoretical worries about fixing upon a 'starting point' ironically may present practical hurdles in the investigation of commodity dynamics. At a very basic level, 'where' should research begin? One of our own projects, for example, commenced with an appraisal of retail sites in the home furnishings 'chain'. A subset of design-oriented retailers were identified, from which suppliers were traced. Consumer interviewees were sought via networks which originated from retailers. Not surprisingly, the apparent scale and complexity of networks and connections increased, and additional actors and sites came into view as the research proceeded. For example, new 'lifestyle' and home design magazines often had the effect of 'leading' retailers and consumers by promoting particular designers and/or styles. Our point is that while researchers might be acutely aware of chain complexities – as well as of 'leakages' between commodity chains (Leslie and Reimer, 1999) – it also would be productive to reflect in more detail upon how different starting points might inflect emergent understandings. Starting points often become politically important, as Raghuram's (Chapter 6, this volume) critique of the simplistic positioning of 'Third World Women' in clothing commodity chain stories demonstrates.

Analytically privileging conceptualizations of networks and circuits over single-stranded chains presents a further methodological challenge. As Crang (2000) has pondered, how do we 'do the whole of geography'? How can we think about, follow and embrace all connections, all of the time? Our own hope is that responses will lie in the continuing pursuit of empirically grounded as well as theoretically and politically informed research at a range of spatial scales – perhaps not always requiring a focus on globalized connections. For the moment, however, we will leave such questions as continuing prompts for future research. We would like to end with a note of thanks to our contributors for their continuing interest and support throughout the production of the book, as well as to participants in the 1998 seminar at the University of Manchester, to which the 'origins' of *Geographies of Commodity Chains* can be traced.

References

Abrams, F. (2002) The long walk. *The Guardian*, 26 June.

Appadurai, A. (1986) Introduction: commodities and the politics of value. In Appadurai, A. (ed.), *The Social Life of Things: Commodities in Cultural Perspective*. Cambridge: Cambridge University Press, pp. 3–63.

Appelbaum, R. and Gereffi, G. (1994) Power and profits on the apparel commodity chain. In Bonacich, E., Cheng, L., Chincilla, N., Hamilton, N. and Ong, P. (eds), *Global Production: The Apparel Industry in the Pacific Rim*. Philadelphia: Temple University Press, pp. 42–62.

Barnes, T. J. (1995) Political economy I: 'the culture, stupid'. *Progress In Human Geography*, 19, pp. 423–31.

Bell, D. and Valentine, G. (1997) *Consuming Geographies: We Are Where We Eat*. London and New York: Routledge.

Busch, L. and Juska, A. (1994) The production of knowledge and the production of commodities: the case of rapeseed technoscience. *Rural Sociology*, 59, pp. 581–97.

Busch, L. and Juska, A. (1997) Beyond political economy: actor networks and the globalization of agriculture. *Review of International Political Economy*, 4, pp. 688–708.

Carrier, J. (1990) Reconciling commodities and personal relations in industrial society. *Theory and Society*, 19, pp. 579–98.

Castree, N. (2001) Commodity fetishism, geographical imaginations and imaginative geographies. *Environment and Planning A*, 33, pp. 1519–25.

Cook, I. and Crang, P. (1996) The world on a plate: culinary culture, displacement and geographical knowledges. *Journal of Material Culture*, 1, pp. 131–53.

Cook, I., Crang, P. and Thorpe, M. (1996) Amos Gitai's Ananas: commodity systems, documentary filmmaking and new geographies of food. Paper presented at the IBG/RGS Annual Conference, Glasgow University, January.

Crang, P. (1996) Displacement, consumption and identity. *Environment and Planning A*, 28, pp. 47–67.

Crang, P. (1997) Cultural turns and the (re)constitution of economic geography. In Lee, R. and Wills, J. (eds), *Geographies of Economies*. London: Arnold, pp. 3–15.

Crang, P. (2000) Discussant's comments on the 'Desiring commodities' session, Annual Meeting of the *Association of American Geographers*, Pittsburgh, April.

Dicken, P., Kelly, P. F., Olds, K. and Yeung, H. W.-C. (2001) Chains and networks, territories and scales: towards a relational framework for analysing the global economy. *Global Networks*, 1, pp. 89–112.

du Gay, P. and Pryke, M. (2002) *Cultural Economy: Cultural Analysis and Commercial Life*. London: Sage.

Fine, B. and Leopold, E. (1993) *The World of Consumption*. London: Routledge.

Friedland, W., Barton, A. and Thomas, R. (1981) *Manufacturing Green Gold*. New York: Cambridge University Press.

Friedmann, H. (1993) The political economy of food: a global crisis. *New Left Review*, 197, pp. 29–57.

Friedmann, H. and McMichael, P. (1989) Agriculture and the state system: the rise and decline of national agricultures, 1870 to the present. *Sociologia Ruralis*, 29, pp. 93–117.

Gereffi, G. (1994a) Capitalism, development and global commodity chains. In Sklair, L. (ed.), *Capitalism and Development*. London: Routledge, pp. 211–31.

Gereffi, G. (1994b) The organization of buyer-driven global commodity chains: how US retailers shape overseas production networks. In Gereffi, G. and Korzeniewicz, M. (eds), *Commodity Chains and Global Capitalism*. Westport, Connecticut: Greenwood Press, pp. 93–122.

Gereffi, G., Korzeniewicz, M. and Korzeniewicz, R. (1994) Introduction: global commodity chains. In Gereffi, G. and Korzeniewicz, M. (eds), *Commodity Chains and Global Capitalism*. Westport, Connecticut: Greenwood Press, pp. 1–14.

Hartwick, E. (1998) Geographies of consumption: a commodity-chain approach. *Environment and Planning D: Society and Space*, 16, pp. 423–37.

Hartwick, E. (2000) 'Towards a geographical politics of consumption'. *Environment and Planning A*, 32, pp. 1177–92.

Harvey, D. (1990) Between space and time: reflections on the geographical imagination. *Annals of the Association of American Geographers*, 80, pp. 418–34.

Hopkins, T. K. and Wallerstein, I. (1986) Commodity chains in the world economy prior to 1900. *Review*, 10, pp. 157–70.

Howes, D. (ed.) (1996) *Cross-Cultural Consumption: Global Markets, Local Realities*. London and New York: Routledge.

Hughes, A. (2000) Retailers, knowledges and changing commodity networks: the case of the cut flower trade. *Geoforum*, 31, pp. 175–90.

Jackson, P. (1999) Commodity cultures: the traffic in things. *Transactions of the Institute of British Geographers*, 24, pp. 95–108.

Jackson, P. (2002a) Commercial cultures: transcending the cultural and the economic. *Progress in Human Geography*, 26, pp. 3–18.

Jackson, P. (2002b) Consumption in a globalising world. In Johnston, R. J., Taylor, P. J. and Watts, M. J. (eds), *Geographies of Global Change: Remapping the World*. Second edition. Oxford: Blackwell, pp. 283–95.

Jackson, P. and Taylor, J. (1996) Geography and the cultural politics of advertising. *Progress in Human Geography*, 20, pp. 356–71.

Kopytoff, I. (1986) The cultural biography of things: commoditization as process. In Appadurai, A. (ed.), *The Social Life of Things: Commodities in Cultural Perspective*. Cambridge: Cambridge University Press, pp. 64–91.

Latour, B. (1991) Technology is society made durable. In Law, J. (ed.), *A Sociology of Monsters: Essays on Power, Technology and Domination*. London and New York: Routledge.

Latour, B. (1993) *We Have Never Been Modern*. Cambridge: Harvard University Press.

Leslie, D. and Reimer, S. (1999) Spatialising commodity chains. *Progress in Human Geography*, 23, pp. 401–20.

McDowell, L. (1997) *Capital culture: Gender at Work in the City*. Oxford: Blackwell.

McMichael, P. (1998) Development and structural adjustment. In Miller, D. and Carrier, J. G. (eds), *Virtualism: A New Political Economy*. Oxford and New York: Berg, pp. 95–116.

McRobbie, A. (1997) Bridging the gap: feminism, fashion and consumption. *Feminist Review*, 55, pp. 73–89.

Marsden, T., Flynn, A. and Harrison, M. (2000) *Consuming Interests: The Social Provision of Foods*. London: UCL Press.

May, J. (1996) A little taste of something more exotic: the imaginative geographies of everyday life. *Geography*, 81, pp. 57–64.

Miller, D. (1997) *Capitalism: An Ethnographic Approach*. Oxford and New York: Berg.

Miller, D. (ed.) (1998) *Material Cultures: Why Some Things Matter*. London: UCL Press.

Murdoch, J. (1997) Towards a geography of heterogeneous associations. *Progress in Human Geography*, 21, pp. 321–37.

Mutersbaugh, T. (2002) The number is the beast: a political economy of organic-coffee certification and producer unionism. *Environment and Planning A*, 34, pp. 1165–84.

Pile, S. and Thrift, N. (1995) Mapping the subject. In Pile, S. and Thrift, N. (eds), *Mapping the Subject: Geographies of Cultural Transformation*. London: Routledge, pp. 13–51.

Powell, W. and Smith-Doerr, L. (1994) Networks and economic life. In Smelser, N. and Swedberg, R. (eds), *The Handbook of Economic Sociology*. New Jersey and New York: Princeton University Press and Russell Sage Foundation, pp. 368–402.

Raikes, P., Jensen, M. F. and Ponte, S. (2000) Global commodity chain analysis and the French filière approach. *Economy and Society*, 29, pp. 390–417.

Raynolds, L. (1997) Restructuring national agriculture, agro-food trade, and agrarian livelihoods in the Caribbean. In Goodman, D. and Watts, M. J. (eds), *Globalising Food: Agrarian Questions and Global Restructuring*. London and New York: Routledge, pp. 119–32.

Sayer, A. (1994) Cultural studies and 'the economy, stupid'. *Environment and Planning D: Society and Space*, 12, pp. 635–8.

Schoenberger, E. (1997) *The Cultural Crisis of the Firm*. Oxford: Blackwell.

Thrift, N. and Olds, K. (1996) Refiguring the economic in economic geography. *Progress in Human Geography*, 20, pp. 311–37.

Watts, M. J. (1992) Living under contract: work, production politics, and the manufacture of discontent in a peasant society. In Pred, A. and Watts, M. J. (eds), *Reworking Modernity: Capitalisms and Symbolic Discontent*. New Brunswick, NJ: Rutgers University Press, pp. 65–105.

Watts, M. J. (1999) Commodities. In Cloke, P., Crang, P. and Goodwin, M. (eds), *Introducing Human Geographies*. London: Routledge, pp. 305–15.

Watts, M. J. and Goodman, D. (1997) Agrarian questions: global appetite, local metabolism: nature, culture, and industry in *fin-de-siecle* agro-food systems. In Goodman, D. and Watts, M. J. (eds), *Globalising Food: Agrarian Questions and Global Restructuring*. London and New York: Routledge, pp. 1–32.

Weiss, B. (1996) Coffee breaks and coffee connections: the lived experience of a commodity in Tanzanian and European worlds. In Howes, D. (ed.), *Cross-cultural Consumption: Global Markets, Local Realities*. London and New York: Routledge, pp. 93–105.

Whatmore, S. and Thorne, L. (1997) Nourishing networks: alternative geographies of food. In Goodman, D. and Watts, M. J. (eds), *Globalising Food: Agrarian Questions and Global Restructuring*. London and New York: Routledge, pp. 287–304.

Part I

Commodity chains, networks and filières

1 From farm to supermarket

The trade in fresh horticultural produce from sub-Saharan Africa to the United Kingdom

Hazel R. Barrett, Angela W. Browne and Brian W. Ilbery

The new global commodity network in fresh horticultural products

In the last 20 years the trade in high value foods (HVFs) such as dairy items, shrimp and fresh horticultural products has become increasingly globalized. HVFs account for over 5 per cent of global commodity trade, one-third of which comes from developing countries (Goodman and Watts, 1997). Just five 'newly-agriculturalising countries' (Friedmann, 1993) are responsible for over 40 per cent of HVFs exported from developing countries. Countries such as Kenya, Brazil and Mexico are among the main world producers of HVFs, most of which are destined for developed world markets in North America and the European Union (EU). Increased consumer demand for year-round fresh horticultural products including highly perishable fruit and vegetables such as strawberries, mangetout and green beans has led to a huge growth in imports of these commodities into the EU. Between 1989 and 1997, for example, imports of fresh and chilled leguminous vegetables increased by 133 per cent with a value of €134 million. Three-quarters of imports came from sub-Saharan Africa, with Kenya supplying over a third of EU imports of leguminous vegetables and Zimbabwe accounting for an additional 10 per cent (Stevens and Kennan, 2000).

The growing market for HVFs in the United Kingdom has been stimulated by two interrelated trends. First, changing consumer tastes have been characterized by an increasing demand for healthy, ethically produced, high-quality food which is available year-round as a 'convenience' product (Barrett *et al.*, 1999; Bell and Valentine, 1997; Ilbery *et al.*, 1999; Miele, 2001; Ward, 1997). Glennie and Thrift (1996) suggest that a process of cultural fragmentation and segmentation has occurred whereby the 'new consumer' emphasizes quality and choice over quantity and price. This is particularly the case with respect to fresh horticultural products. Sales of speciality vegetables and pre-washed salads in the United Kingdom increased by 21 and 34 per cent respectively in the period 1993–96 (Fearne and Hughes, 1998). Second, corporate retailers have responded robustly to the demands of the 'new consumer'. Supermarkets now sell over 70 per cent of fresh horticultural products and their market share is increasing. They offer year-round supplies sourced from different parts of the world and insist on complete product traceability and strict health and hygiene standards. Corporate retailers effectively regulate quality standards in a private interest re-regulation of the food sector (Marsden *et al.*, 2000). They are both affected by changing consumer demands and at the same time seek to manipulate them.

The production of fresh horticultural products for export has grown rapidly in a number of sub-Saharan African countries over the last decade for several reasons (Barrett and Browne, 1996). First, changes in the international trading environment such as EU trade preferences under the Lome Convention, EU bilateral agreements and the Generalized System of Preferences, have given sub-Saharan African countries a competitive advantage over non-preferred exporters (Stevens and Kennan, 2000). Second, export-led growth and agricultural diversification into HVFs were facilitated in sub-Saharan Africa in the 1990s by liberal economic restructuring policies supported by the Bretton Woods Institutions in response to debt and rising poverty (Barrett and Browne, 1996). The favourable international market for HVFs and opportunities for new product development and value-adding activities (Jaffee and Morton, 1995) encouraged investment in fresh horticultural products for export. Third, technological improvements in storage and transportation have meant that produce can be on the shelves of UK supermarkets within 48 hours of harvesting.

African countries exporting fresh horticultural products to the EU appear to form a fairly stable group. As Table 1.1 shows, the key players are Kenya, Zambia and Zimbabwe, with countries such as Cameroon, Morocco, Tanzania, The Gambia and Tunisia competing hard. It is a fiercely aggressive market, with producers having to innovate and improve efficiency to maintain or increase market share. Market linkages, together with knowledge and information, are key elements of competitive advantage in this global commodity network, which is increasingly founded on the rapid interchange of data and information (Fearne and Hughes, 1998) and flexible contractual arrangements with corporate retailers. The result has been the development of a new technologically driven global commodity network, in which the speed of delivery and product quality are paramount. It is a network linking farms producing fruit and vegetables in sub-Saharan Africa through a 'cool-chain' to supermarkets in the United Kingdom which ultimately supply consumers, who in turn influence production by their changing tastes and demands. This new global network of trade in fresh horticultural products is the focus of this chapter.

The chapter reviews various approaches to the study of global commodity linkages and suggests that not one adequately conceptualizes this new international trade. The network approach provides a good basis as it incorporates business and economic dimensions of the international trade in fresh horticultural products. However it fails to give full consideration to innovation and knowledge circuits that our case study identifies as vital for success. We therefore suggest that the notion of a cellular network – in which business and product linkages are connected by overlapping circuits of knowledge and innovation – provides a more appropriate conceptualization. Our case study of the trade in fresh vegetables from Kenya and The Gambia to corporate retailers in the United Kingdom

Table 1.1 African countries supplying fresh vegetable exports to the EU (1997)

Exports	Worth more than €1 million	Worth more than 5% of EU market
Peas	Kenya Zambia Zimbabwe	Kenya Zambia Zimbabwe
Beans	Burkina Faso Egypt The Gambia Kenya Mali Morocco Senegal Zambia Zimbabwe	Burkina Faso Egypt Ethiopia Kenya Morocco Senegal
Other vegetables	Egypt Ghana Kenya Morocco South Africa Tunisia Uganda Zambia Zimbabwe	Kenya Morocco

Source: Adapted from Stevens and Kennan, 2000, appendix 1, pp. 35–6.

(see also Barrett *et al.*, 1997) demonstrates that success is dependent on the level of engagement and embeddedness by actors and sites in linkages and circuits identified in the cellular network.

Conceptualizing global commodity networks

The concept of a commodity chain, circuit or network examines the links and progressive movement of a commodity through the sequential phases of production, distribution and consumption (Hughes, 2000; Leslie and Reimer, 1999). Commodity chains '…consist of significant production, distribution and consumption nodes, and the connecting links between them, together with social, cultural and natural conditions involved in commodity movements' (Hartwick, 1998, 425).

In this way, social and natural conditions at the production end of the chain are integrated with the cultural (i.e. signifying and representational) effects at the consumption node (Hartwick, 1998). This also suggests that commodity chains consist of both 'vertical' and 'horizontal' dimensions. While the former include production, distribution and retailing across different commodities, the latter act as connecting nodes for commodity movements and thus include such dimensions as places, gender, class and ethnicity. As Hartwick (1998) asserts, the radiating effects of the consumption node, which is connected to the social and natural conditions at the production node by complex intersections at various interme- diating and terminal points, produces a simple model of commodity movement which integrates vertical and horizontal dimensions.

Over time, geographers and other academic researchers have developed dif- ferent approaches to the study of commodity linkages. Three in particular have informed our research.

Commodity chains

Deriving from the political economy paradigm, this approach depicts a unilinear chain whereby commodities are produced in 'peripheral' regions of the global economy for retail and consumption in the 'core' (Hughes, 2000). Commodity chain analysis focuses on actors' connections between core and periphery and their spatial reach. As chains extend in length, so socio-technical complexity increases (Murdoch, 2000). All connections are deemed to be determined by power rela- tionships. Such power may be producer-driven (as in automobiles) or buyer-driven (as in most food commodities) (Gereffi, 1995). The latter is the focus of this chapter, which emphasizes the power of agribusiness/retail capital in defining the shape of fresh horticultural commodity chains in Kenya and The Gambia.

Lowe and Wrigley (1996) discuss how large retailers have managed and dom- inated the food commodity chain, working closely with suppliers on product design, process innovation, quality monitoring and distribution logistics. However, as Murdoch (2000, 410) explains 'While the commodity chain approach clearly highlights the complex socio-natural composition of contem- porary food networks, it tends to see these as always and necessarily composed in line with particular relations of power.'

Both Arce and Marsden (1993) and Leslie and Reimer (1999) have criticized the commodity chain approach for being reductionist in its focus on vertical dimensions of the chain while largely ignoring horizontal aspects relating to place and space, such as class, ethnicity and gender. The approach is unidirectional and deterministic because it often takes retailing and consumption as unproblematic starting points from which to examine exploitation in the productive system.

In response to such criticisms, Fine (1994) advocated a 'systems of provision' approach to provide a 'more balanced treatment of the relationship between pro- duction and consumption' (Leslie and Reimer, 1999, 405). Each commodity possesses a unique 'system of provision', suggesting a need to study the vertical dimensions of particular commodities or groups of commodities and the mate- rial culture surrounding these dimensions (Fine, 1994; Fine and Leopold, 1993). While incorporating a more dynamic consideration of consumer behaviour and avoiding the core–periphery dualism, Fine's work has been criticized for its inability to deal with the complexity of consumption practices, which are treated as simple outcomes of activities associated with the provision of food. As Lockie and Kitto (2000, 5) explain '…activities involved in the production, processing, distribution and retailing of food, in combination with socioeconomic variables, are treated as *determinants* of consumption practices, despite the lack of evidence of any causal relationship between any of these activities or variables and food consumption practice'.

There is little attempt to examine the complexity of consumption practices. This has led to a shift in interest by some academics away from chains and towards circuits.

Commodity circuits

Embedded within a broader literature on commodity cultures, a commodity circuits approach treats 'the movement of commodities through phases of production, distribution and consumption as a non-linear circuit, rather than a linear chain' (Hughes, 2000, 177). It is recognized that production is consumer driven and that producers are also consumers (Cook and Crang, 1996). Whereas commodity chain analysis may privilege one site over others, explanation in commodity circuits lies in a combination of different processes and posits a need to examine the dense web of interactions between all sites. As Leslie and Reimer (1999, 406) explain, 'Rather than assuming that consumers have little knowledge of commodities, their quality and geographical origins, commodity circuits both construct and are reconstructed by consumer knowledge.'

The circuits of culture approach seeks to arrive at a more contextual understanding of the meanings attached to goods in different times, places and phases of commodity circulation. It thus embraces horizontal dimensions and focuses on 'the complex ways in which goods are displaced from one site to another, and the ways in which various cultural knowledges inform, and are informed by, this process of displacement' (Hughes, 2000, 177).

However, by emphasizing commodities as complex cultural forms (Jackson, 1999) and advocating a shift towards sites of consumption, the 'circuits of culture' approach has been criticized for downplaying the significance of power and political economy and overemphasizing the role of consumers. Thus aspects of a circuits approach should be reconnected to the competitive pressures identified by commodity chains – and a useful metaphor for rethinking this relationship between production and consumption is the network.

Commodity networks

In an attempt to avoid privileging any one-dimension of commodity circulation, a network approach focuses on 'how different kinds of nodes (people, firms, states, places and organizations) are connected to one another in complex and multi-stranded ways' (Hughes, 2000, 178). At a general level, commodity networks begin to grip the pattern of 'webs of interdependence' that exist between different actors in the rural economy (Ilbery and Kneafsey, 1999; O'Neill and Whatmore, 2000). Webs of interdependence 'not only connect firms through vertical commodity exchange relationships but also bind together additional agents through the multi-directional flows of information and materials that variously support these exchange relationships' (Hughes, 2000, 178). Thus the privileging of one sphere of commodity circulation is avoided, as the network is extended to include other sites such as research and development, non-governmental organizations and consumer groups. The approach views relations between producers, institutions, retailers and consumers as the outcome of complex flows between a range of interconnected actors that have become enrolled in the network. Crucially, the network integrates both vertical and horizontal dimensions of commodity movement (Murdoch, 2000; Whatmore and Thorne, 1997).

A good example of the benefits of adopting a 'network' rather than a 'chain' approach to commodity movements can be seen in recent campaigns for ethical trade (Blowfield, 1999; Hartwick, 2000; Johns and Vural, 2000). While retailers may have the most power and a growing influence on global commodity networks, they are susceptible to campaigns by consumer groups, the media and non-governmental organizations, who become additional agents in the network. Campaigns have focused on many different sectors of the global economy with the fashion industry being one of the most visible (see Crewe, Chapter 10, this volume). Similar, but lower-profile campaigns have been directed towards the global trade in food products. Comparable issues, such as the increasing power of Western retailers, the poor working conditions of workers and out growers, gender inequality, the disadvantages faced by small-scale farmers as well as environmental issues have all been raised by campaigners and the media with reference to the global food supply network. Christian Aid has campaigned against the unfair trading activities of supermarkets (ranking them according to the extent of their ethical practices) and the poor working conditions of many producers. Likewise, Twin Trading, a campaigning and trading company, has criticized the trading practices of large-scale retailers and as a consequence has deliberately sought to support small-scale producers of products such as tea and coffee. Global commodity networks have thus been expanded, diverted or repositioned by 'campaigning' agents. As a consequence of such developments, retailers and other businesses are under pressure to act more responsibly and ethically and have introduced codes of practice/conduct that set minimum standards, usually at the production node of the network. Unfortunately, codes of practice, monitoring and auditing often are established by the most powerful actors in the 'core' and imposed on workers in the 'periphery' (Blowfield, 1999) and so there is a need for greater worker participation in their development and implementation.

Different approaches to commodity movements treat knowledge and information in different ways. As Hughes (2000, 179) demonstrates '...knowledge plays a vital part in re-shaping the morphology of a commodity network and defining the flows within it is central to an understanding of how retailers shape...and influence the process of production'.

While knowledge often is understood as fixed and 'one way' in the commodity chain approach, circuit and network analyses view it as multifaceted, circulating between producers, retailers and consumers. Retailers' use of knowledge can be seen as drawing upon and transforming ideas developed in other nodes such as production, consumption, and research and development. These overlapping circuits or cells ensure innovation and success within the commodity network.

Trade in fresh horticultural products represents a special system of provision, which we conceptualize in this chapter as a global commodity cellular network. The term network is preferred because it enables us to explore the complex set of interrelationships between key actors and sites. It also permits an examination of both vertical and horizontal dimensions of commodity movement and aspects of consumer culture, while retaining ideas associated with corporate retailer power in the commodity chain approach. Crucially, a focus on networks allows us to highlight flows of information and knowledge between many interconnected actors enrolled in the fresh horticultural commodity network. The production of fresh fruit and vegetables in Kenya and The Gambia is not simply 'determined' by corporate retailers in the United Kingdom, but rather is negotiated by a variety of agents at different nodes in the network. A cellular network approach thus fuses aspects of earlier materialist commodity chain analyses with more recent writing on geographies of consumption (Barrett *et al.*, 1999; Crewe, 2000; Hughes, 2000).

The chapter focuses on the analysis of several key factors and dimensions surrounding the movement of fresh horticultural commodities from production sites in Kenya and The Gambia to supermarket shelves in the United Kingdom. It is recognized that the views of the final consumers are not represented here; the 'black box' of consumption (Lockie and Kitto, 2000) deserves and awaits more detailed attention.

The fresh horticultural commodity network linking sub-Saharan Africa to the United Kingdom

The example of fresh horticultural exports from Kenya and The Gambia to the United Kingdom illustrates the complexity of global networks and the ways in which agribusiness and retail capital converge to create a complex system of production, export, retail and consumption activities across an extended spatial reach. Both countries are involved in the global trade in horticultural production, with Kenya being one of the 'newly-agriculturalising countries' identified by Friedmann (1993). Although Kenya in particular exports a wide range of horticultural products, our discussion focuses on high-value fresh vegetables. These are of considerable interest in the global food system because they must be moved swiftly from point of production to consumption and are subject to constant change in retailers' requirements. As a fresh food product that is only minimally transformed between farm and supermarket, high-value vegetables must be safe to eat and appealing to consumers. The network therefore relies on efficient and rapid interchange of knowledge, information and ideas, as well as speed in transit of the product.

The cellular network approach adopted here considers sites of retailing (and to a limited extent consumption), import, export and production within a framework that investigates actors (nodes) and the movements between them, not as unidirectional but as multi-directional exchange relationships (represented diagrammatically in Figure 1.1).

As will be shown, Kenya has many more actors in this commodity network due to its longer involvement in the international trade in horticultural commodities

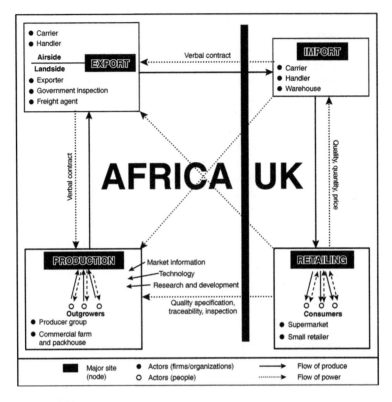

Figure 1.1 The HVF commodity network linking sub-Saharan Africa to the United
   Kingdom.

Source: Authors' fieldwork.

and its much greater variety of products and sourcing methods (Jaffee, 1995).
As a more recent entrant into this trade and with only a few production sites, The
Gambia has a simpler network with fewer vertical and horizontal connections.
For both countries, the discussion demonstrates that while produce does flow
one way within a commodity chain that links production spaces in the periphery
with consumption sites in the core, flows of information, knowledge and power
are multifaceted, forming a complex network based on overlapping circuits of
knowledge and innovation.

Sites of retailing and consumption

Corporate retailers dominate the sale of imported fresh horticultural products in
the United Kingdom. Other retailing sites for produce originating in sub-
Saharan Africa are small supermarkets, high-street greengrocers' shops and
market stalls. Produce reaches these retailing nodes by two different, but often
interlinked, marketing chains, labelled the 'supermarket chain' and the 'wholesale

chain'; both are efficient in delivering to the retailers produce of the quality demanded by the stores and customers (Barrett *et al.*, 1999).

The two retail sites (supermarkets and smaller retailers) have different priorities in their retailing strategy and, for the most part, a different consumer base. For supermarkets, fresh produce is considered to be a 'window' on their store, setting a marker for quality and signifying their target market through a mix of volume and luxury lines and high quality packaging and presentation. High-value vegetables are presented as healthy, fresh and convenient, being offered for sale ready trimmed and pre-packed, sometimes within a mixed pack ready to be microwaved or stir-fried. By contrast, small retailers, particularly those selling in retail markets and ethnically diverse urban areas, are more concerned with offering produce that is competitively priced and of good quality. Traders interviewed for this research in wholesale markets in London and Birmingham confirmed that their retail customers were more concerned with value for money and with the flavour of produce than with its presentation or packaging (Barrett *et al.*, 1997). Many of these retailers trade in speciality produce for Asian and other ethnic cuisine as well as volume lines. Their customers form a loyal and established group.

Although this chapter does not specifically investigate consumption sites, retailers are under pressure to address the demands and ethical interests of the 'new consumer', who is concerned about issues such as fair trade, worker exploitation and environmental impacts within production sites in the developing world (Gabriel and Lang, 1995; Wells and Jetter, 1991). The ethical trading movement has had a large impact on sourcing of HVFs by supermarkets. Some HVFs, for example, prawns and asparagus were specifically singled out in Christian Aid's (1996) campaigning report, *The Global Supermarket*, which identified worker exploitation among suppliers to major supermarkets. The report elicited prompt responsive action. A year after the report's publication, seven out of the top ten supermarket chains had adopted ethical policies (Christian Aid, 1997) and since that time supermarkets have taken steps to incorporate systems to check and monitor labour conditions and, where relevant, environmental impacts throughout their supply network.

All the major UK-owned supermarkets are members of the UK's Ethical Trading Initiative (ETI), an independent alliance of leading retailers, Non-Governmental Organizations (NGOs) and the international trade union movement. ETI was set up in 1998 to promote the adoption and promulgation of better labour conditions around the world and has developed codes of conduct and practice to monitor this. Within this forum supermarkets are actively involved in the implementation of the codes, particularly those laying out minimum labour standards. They also cooperate with pilot projects to develop independent verification methodologies for these codes. As a condition of membership of the ETI retailers must undertake to apply the ETI base code and report annually on their progress in supply chain monitoring for compliance with this code (ETI, 2000, 2001). Companies recognize that audit methodologies tend to find 'easy to spot' contraventions, such as health and safety standards, rather than sensitive labour issues such as discrimination or abuses of workers' rights, and that more independence and transparency are needed in monitoring methodologies (ETI, 2001). Even so, supermarkets now claim to be implementing 'ethical' or 'socially responsible' sourcing and are said to be taking ethical issues from the 'niche' to the 'mainstream' marketplace (Browne *et al.*, 2000). This has great importance for production sites within the HVF commodity network because it requires codes of practice, monitoring and auditing, informed by the interests of many different actors from the core, to be met and understood by producers in the periphery (Blowfield, 1999).

Supermarkets have a direct interest in knowing exactly where produce originates due to their need to show 'due diligence' in ensuring food safety. The 'due diligence' clause within the UK Food Safety Act (1990) gives retailers legal responsibility for the safety of all food products that they sell, and they can be charged under this Act for failing to take all reasonable precautions to ensure this (Doel, 1996). Consequently, as Doel (1996, 62) explains, 'in some cases, ardent "policing" by retailers has filtered up the supply chain', requiring close links with trusted suppliers and strict regimes of vetting. For retailers, both food safety requirements (regulated by law) and ethical trading issues (unregulated, but validated by internal audit) are powerful determinants of which farms they buy from. Inspection and vetting are crucial to the integrity of their claims, requiring intensive interaction between retailers and their suppliers.

The inspection of farms by technical, marketing and audit personnel from supermarkets is conducted periodically, with all stages of production, processing and record keeping being scrutinized. Any fault in practice or paperwork can threaten farmers' supply agreements with that supermarket or its agents. Although reports of 'contract' termination are rare, the inspection process nevertheless ensures absolute compliance with UK supermarket demands and imposes control over all aspects of HVF production. At all times, farmers must respond immediately to demands from supermarkets. In one episode witnessed during our fieldwork, a list of the ages of all 5,000 employees on a Kenyan farm was requested by a major UK retailer (by fax the same day) in response to criticism in a press report that the company allowed its suppliers to employ child labour. In this case, the action and reaction were informed by information flows within and between the core and periphery, mediated through the UK press and founded upon western cultural assumptions of ethical business behaviour. In fact, the producer employs nobody under 16, in common with other commercial farmers in Kenya who subscribe to codes of practice established in the 'core'. In addition to such one-off requests, another intersection between core and periphery is the daily communication from supermarkets advising on labelling for price changes, special offers and other promotional tactics. Packs of mangetout with stickers proclaiming 'only 99p' and on green beans with '50p off', supplied by retailers in their corporate style, were awaiting despatch from one farm visited for this research. Likewise, each supermarket supplies its own packaging materials, including plastic trays, cellophane and bar code stickers, and any required adjustment in product appearance is conveyed daily to suppliers. Plate 1.1 shows runner beans being packed for a UK supermarket at an on-farm facility in Kenya.

Supermarkets' concerns for food safety and high quality of product freshness and appearance require considerable investment in handling, packaging and

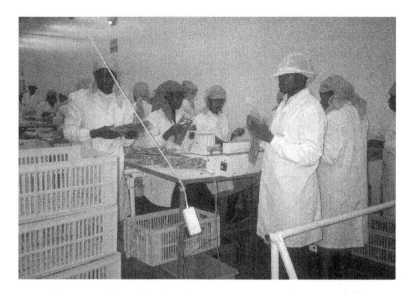

Plate 1.1 Runner beans being weighed and packed at an on-farm packaging facility in Kenya.

Source: H. R. Barrett.

transportation infrastructure at production sites and at other major nodes in the commodity network, as the following sections will make clear. Corporate retailers in the United Kingdom therefore hold considerable and increasing power over production sites. Requirements for consistency, quality and traceability (see Barrett *et al.*, 1999) exert both a direct and indirect impact on farmers in Kenya and The Gambia. In return, retailers receive a year-round flow of high quality produce.

Sites of import

As Figure 1.1 illustrates, supermarkets are supplied with HVF products by importing companies. These actors provide the vital link in the network between farmers and exporters in Africa and retailers in the United Kingdom. Although produce flows from importer to supermarket, there is a reverse flow of power due to the nature of the contractual arrangements between the two. Importing companies must find suppliers, ensure that technical, quality and transportation standards are met and deliver exactly the quantity and quality of each product line requested by supermarkets at a price that is acceptable.

Supermarkets engage with importers through what are often described as 'hand-shake agreements', which are contractual relations based on trust rather than legally binding written contracts. Practices in the UK manufactured foods sector identified by Doel (1999) – where commercial contracts between retailers and UK suppliers of own-label foods are extremely rare and ties are based on agreements to supply, without fixed time periods or guarantees – are mirrored across international space. Our study also echoes Doel's (1996) finding that failure by UK food manufacturers to deliver can result in termination of agreements. Importers unable to deliver on agreed quantity, quality or price can risk losing supply agreements at short notice and without compensation, and breaking the 'agreement' is the prerogative of the retailer. As Doel (1999, 76) explains, '…retailers do not *need* contracts because the power differentials they enjoy render suppliers compliant and effectively remove planning problems from their domain'.

Most importers deal with only a few large-scale HVF suppliers within each country and sometimes – as with one importer sourcing from The Gambia – with only one dedicated supplier. (In this case, the importer and the Gambian farm are part of the same commercial group.) Importers maintain overseas personnel in their primary source country and like supermarkets, conduct overseas visits to farms, pack houses and warehouses to undertake quality inspections.

Importers continually must be searching for new suppliers to ensure year-round supplies and meet new demands. Unpredictable events such as climatic irregularities or political upheaval may interrupt supplies from a particular source. Following a successful military coup d'etat in The Gambia in July 1994, many air carriers withdrew their services and tourism slumped, affecting the availability of freight space. Exports of fresh horticultural products were cut off virtually overnight and without warning. More predictable disruptions include inefficiencies by airlines and their handling agents, either at the point of export or while in-transit. Perishable fruit and vegetables must be transported speedily along a 'cool chain'. Thus securing freight-space, speeding produce through cargo handlers once consignments land in the United Kingdom, and meeting supermarkets' tight delivery schedules all impact upon the ability of importers to retain their contracts in a highly competitive market.

Sites of export

For perishable horticultural products, the sites of export are the major international airports, Jomo Kenyatta International Airport (Nairobi) in Kenya and Banjul International Airport in The Gambia. Many actors in the commodity network converge on these nodes, some 'landside' and others 'airside', as shown in Figure 1.1. As this section will show in the case of Kenya, different actors operate landside and airside, with varied connections and intersections and diverse power and knowledge bases. Breaking through from landside to airside quickly is the key to swift export and transport of perishable produce, but this requires power, efficiency by the relevant actors and good 'contacts'. As a major exporter of horticultural commodities, Kenya's Nairobi airport is a site of intense activity by the many interconnected agencies of this global network.

Three important 'landside' actors at Nairobi Airport are the government regulatory agents, freight agents and exporters. The flow of knowledge and information between these interconnected actors and others in the supply network is essential at this export node. The Kenyan government has strict regulations on the export of horticultural goods to ensure its reputation and credibility as a world-class producer. At the airport, licensed exporters, usually through their freight or handling agents, must acquire a phyto-sanitary certificate following inspection by the Crop Protection Agency of the Ministry of Agriculture, a weight manifest issued by Kenya Airways Handling Ltd (KAHL), and customs documents. The paperwork accompanying horticultural consignments is proof that they meet EU and UK legislation, as well as quality standards required by importers and supermarkets. Actions taken at the airport, which are determined by international regulation, promote Kenya's reputation within this global trading network.

Exporters in Kenya appoint freight agents to buy airfreight space, handle produce at the airport and acquire the necessary paperwork. Approximately 200 Kenyan companies hold export licences. The largest horticultural exporters have vertically integrated their marketing operations by setting up their own export and freight companies which increase their power and control within the global fresh horticultural commodity network. For smaller exporters, the freighting of produce can be problematic. Costs are said to be expensive and freight space is not always available on flights to Europe, even if pre-booked. Smaller agents have no power in their relationships with airlines: their complaint that large exporters 'have contacts' indicates that, even within one activity at this site, some actors are privileged over others.

Handlers and carriers 'airside' are the final agents of importance at the Kenyan export node. Handling and loading by ground staff is a source of concern to some exporters, who see favouritism and 'contacts' as being necessary for airside concessions and direct access on to aircraft. Some of the largest exporters are permitted to operate cold stores as 'airside', where produce is inspected and certified in the warehouse and then loaded directly on to planes. For others not thus privileged, there may be lengthy delays and produce is sometimes 'left standing on the tarmac'. This is a source of irritation to many agents, but emphasizes the importance of connections at this node.

In marked contrast to Kenya, at The Gambia's international airport the number and variety of actors is very few. There is no government inspectorate: quality control and certification are handled by the largest commercial farmers who complete the paperwork for their own consignments. There is only one freight agent at the airport, who buys space on passenger charter flights and during peak season charters cargo planes. The company has developed an international freight centre at the airport where produce is palletized and weighed before being loaded on to planes. Plate 1.2 shows green beans being checked into the 'airside' facility at Banjul International Airport. Although the freight centre operates as 'airside', low levels of airport security blur the distinction between 'landside' and 'airside'. Production, transportation and import into the United Kingdom of fresh horticultural produce from The Gambia are greatly facilitated and power relationships neutralized by the fact that all principal actors are sister companies within one commercial group. Ties of kinship further connect the companies, as separate companies within the group are owned by different family members. The same group also has interests in the Kenyan horticultural export sector, illustrating another dimension to the complex web of vertical and horizontal interconnections in this global network.

Plate 1.2 Green beans being checked into a temperature-controlled pack-house at Banjul International Airport, The Gambia.

Source: H. R. Barrett.

Sites of production

There are many facets of the fresh horticultural commodity network that favour large commercial farms in Kenya and The Gambia over smallholders. Large-scale farmers have resources of knowledge, literacy, technology and infrastructure that permit them to meet supermarkets' requirements and crucially, to provide the documentation to prove it. They draw upon knowledge, expertise and inputs from around the world, including Israeli drip-feed irrigation advisers, UK agronomy graduates and Dutch seed stock. In Plate 1.3, women labourers prepare vegetable seed plugs at a commercial farm in The Gambia.

Large-scale farmers also scan commercial publications and the Internet for up-to-date information on food trends and fashions in the United Kingdom. Some of the largest horticultural farms in Kenya have on-farm research and development facilities, where new products such as 'baby' vegetables or new cultivation techniques such as hydroponics can be tested. In this way, producers hope to maintain their competitive position within the global sourcing strategy of buyers and be ready to respond quickly to new demands. This is how they retain power in an otherwise unequal power relationship.

Large-scale producers are also keenly aware of the quality, presentation and packaging requirements of supermarket customers and for this reason have invested heavily in EU-standard pack houses at farm-sites (in the case of the largest growers in Kenya) or the airports. Vegetables are graded, prepared, packed,

Plate 1.3 Women labourers prepare vegetable seed plugs at a HVF commercial farm in The Gambia.

Source: H. R. Barrett.

labelled and priced, ready for the shelf, according to the specifications of each retail chain that the farm supplies. Like the production sites, pack houses are inspected regularly by supermarket produce buyers and must consistently meet their standards of quality and safety. By making this investment, producers hope to maintain the loyalty of supermarkets, but at the same time supermarkets are dependent on the efficiency of these facilities. The power relationship is thus two-way.

Smallholders can also reach international markets, but through a range of intermediaries in a long supply chain. Many small- and medium-scale farmers in Kenya (though very few in The Gambia) access international markets indirectly by growing on contract to the large producers as out growers (Figure 1.1). Green beans, the main out grower vegetable crop, are grown according to strict schedules determined by the contractor, which aim to ensure a constant supply at the central depot of a uniform and high quality product. Out growers also are used by the two major commercial farms in The Gambia, but only to make up shortfalls in supply (for chillies and aubergines) and under very loose contractual conditions. 'Verbal agreements' are said by out growers to be virtually meaningless, with low farm-gate prices and unreliable collections.

Another route by which smallholders enter the international commodity network is via the 'wholesale chain'. Many thousands of Kenyan farmers sell produce to 'middlemen', agents, large growers or exporters in a web of non-contractual and often unreliable arrangements. The problems faced by smallholders in this risky market are many. Low prices, agents not appearing to collect 'agreed' consignments, and prices being reduced for produce described as 'below quality' all illustrate the relative powerlessness of smallholders in this network. On the other hand, agents and exporters complain that smallholders either renege on verbal agreements and/or sell their produce to non-appointed agents who offer spot cash at the farm-gate. At these production sites, therefore, incomplete market knowledge and unreliable commercial arrangements render engagement by smallholders with the international commodity network a risky business.

A new approach to the global HVF network

This chapter has focused on a new network, which has been established for the global trade in fresh horticultural produce, where perishable products must be handled, packaged and transported with speed along a temperature-controlled 'cool chain'. This requires high levels of investment, which necessarily biases production sites towards large-scale commercial growers with whom other actors and sites in the network liaise. As the case study has demonstrated, the nature of arrangements between corporate retailers, importers and exporters is very complex.

The fresh horticulture global commodity network functions within and responds quickly to changes in the regulatory framework. The trade operates under national regulations, which incorporate international agreements. In the United Kingdom, the industry is regulated by the 1990 Food Safety Act and 1996 Food Hygiene Directive. Increasingly, corporate retailers in the United Kingdom are taking on the responsibility of interpreting and monitoring this legislation (Marsden *et al.*, 2000). As private-interest self-regulators condoned by the government, retailers have endowed themselves with regulatory power within the network. Both directly and through their intermediaries, retailers are able to impose this power upon growers in sub-Saharan Africa.

Based on capitalist accumulation and comparative advantage, Dolan and Humphry (1999) believe that the international trade in fresh horticulture commodities has many of the characteristics of a buyer-driven global commodity chain. Retailers do not own production facilities: instead, they rely on a complex, tiered network of 'subcontractors' that perform almost all specialized tasks, including growing, processing, packaging and transporting produce. Retailers also are able to respond quickly to changes in the international trading environment and to switch sourcing countries without a loss of investment. Thus corporate retailers command great economic power while taking little financial risk. Although this case study has demonstrated the importance of both regulatory and economic power within the global fresh horticultural commodity network, it also has illustrated that the flow of power is not always unidirectional. Further, complex and interlinking circuits of knowledge and innovation are associated with the various nodes within the network. It is these which explain the success of the industry in Kenya, compared with the limited achievements of The Gambia.

The network approach allows the vertical and horizontal dimensions of many sites – from production and distribution to marketing and retailing – to be brought together as shown in Figure 1.2 which is conceptualized from the Kenyan example.

Consumer, retailer and importer/exporter knowledges, interwoven within 'circuitous geographies of knowledge' (Cook *et al.*, 2000, 252), affect relations between actors and impact upon behaviour at every node within the network. The case study confirms that knowledge circuits are an important factor in the privileging of sites and actors and in the overall 'success' of a commodity network.

Although our research did not include detailed analysis of the site of consumption, it is evident that consumer demand has been an important element within the fresh horticulture network. Within networks of stable structural and power relations, factors associated with the site of consumption become important, especially in terms of market share and economic performance. The ability to predict, manipulate and respond quickly and positively to changing consumer preferences is seen as central to any successful economic enterprise and is particularly the case in the highly competitive supermarket sector. This necessitates the development of new networks of knowledge and innovation around nodes and between actors within the network. As Murdoch (2000, 412) states 'It is now widely believed that economic performance within the global economy is heavily reliant upon the capacity to *innovate* rather that [sic] simply to participate.'

The case study shows that in such a competitive trade as HVF innovation at each node, namely production, export, import and retailing, is essential if actors are to remain competitive and maintain their position in the network. However, each knowledge circuit overlaps with others in the vertical elements of the network, producing a complex web of interdependence between various actors, forming what Murdoch (2000, 415) labels a 'learning economy of innovation'. Well-established sites of production such as large commercial farms in Kenya have a sophisticated learning circuit of innovation and knowledge. This is shown by the introduction of new crops and production technologies as well as investment in processing and packaging facilities. Such investments respond to shifts in consumer demand as perceived and articulated by the supermarkets. Due to this, Kenya has maintained and consolidated its status as a global newly-agriculturalising country (NAC) in terms of high value fresh horticultural exports. 'Newcomer' countries such as The Gambia are placed at a disadvantage. Access to knowledge, innovation and new technology, all part of the circuit of knowledge, is crucial to economic success in the global fresh horticultural commodity network and requires knowledge circuits that can detect and respond quickly to changes occurring in the various nodes of the vertical network.

The two examples demonstrate the importance of both circuits of knowledge and vertical networks for the success of this trade. Figure 1.2 indicates that in the Kenyan case this new network consists of a series of nodes and complex interconnections. However, within the network there is a flow of power, indicating a structural chain which privileges some sites and actors over others. In addition, each node has an associated circuit of knowledge and innovation. The competitive nature of the network is ensured by the overlapping and interdependent nature of knowledge circuits. These allow some actors to respond quickly to changes in consumer demand, the regulatory framework and the trading environment, as is evidenced in the Kenyan example. Thus a cellular network is formed in which

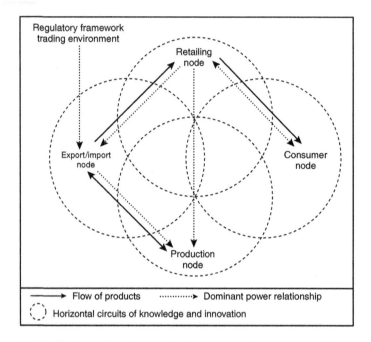

Figure 1.2 The horizontal and vertical dimensions of the new global commodity network for HVFs.

Source: Authors.

connections of circuits of knowledge and innovation are as important as vertical business/product linkages in ensuring economic success, a lesson that farmers in The Gambia and other emerging producer countries must learn.

Acknowledgement

The authors gratefully acknowledge the support for this research from the ESCOR section of DFID (Research grant R6139). The views and opinions expressed in this chapter do not reflect DFID's official policies or practices, but are those of the authors alone.

References

Arce, A. and Marsden, T. (1993) The social construction of international food: a new research agenda. *Economic Geography*, 69, pp. 293–312.

Barrett, H. R. and Browne, A. W. (1996) Export horticultural production in sub-Saharan Africa: the incorporation of The Gambia. *Geography*, 81, pp. 47–56.

Barrett, H., Ilbery, B., Browne, A. and Binns, T. (1999) Globalization and the changing networks of food supply: the importation of fresh horticultural produce from Kenya into the UK. *Transactions of the Institute of British Geographers*, 24, pp. 159–74.

Barrett, H. R., Browne, A. W., Ilbery, B. W., Jackson, G. H. and Binns, T. J. (1997) Prospects for horticultural exports under trade liberalisation in adjusting African economies. Report submitted to Department for International Development (DFID).

Bell, D. and Valentine, G. (1997) *Consuming Geographies: We Are Where We Eat*. London: Routledge.

Blowfield, M. (1999) Ethical trade: a review of developments and issues. *Third World Quarterly*, 20, pp. 753–70.

Browne, A. W., Harris, P. J. C., Hofney-Collins, A. H., Pasiecznik, N. and Wallace, R. R. (2000) Organic production and ethical trade: definition, practice and links. *Food Policy*, 25, pp. 69–89.

Christian Aid (1996) *The Global Supermarket*. London: Christian Aid.

Christian Aid (1997) *Change at the Check-Out? Supermarkets and Ethical Business*. London: Christian Aid.

Cook, I. and Crang, P. (1996) The world on a plate: culinary culture, displacement and geographical knowledges. *Journal of Material Culture*, 1, pp. 131–53.

Cook, I., Crang, P. and Thorpe, M. (2000) 'Have you got the customer's permission?': category management and circuits of knowledge in the UK food business. In Bryson, J. R., Daniels, P. W., Henry, N. and Pollard, J. (eds), *Knowledge, Space, Economy*. London: Routledge, pp. 242–60.

Crewe, L. (2000) Geographies of retailing and consumption. *Progress in Human Geography*, 24, pp. 275–90.

Doel, C. (1996) Market development and organizational change: the case of the food industry. In Wrigley, N. and Lowe, M. (eds), *Retailing, Consumption and Capital*. London: Longman, pp. 48–67.

Doel, C. (1999) Towards a supply-chain community? Insights from governance processes in the food industry. *Environment and Planning A*, 31, pp. 69–85.

Dolan, C., Humphry, J. and Harria-Pascal, C. (1999) Horticulture commodity chains: the impact of the UK market on the African fresh vegetable industry. Working Paper 96, Brighton: Institute of Development Studies.

Ethical Trading Initiative (ETI) (2000) Annual Review 1999/2000 Getting to work on ethical trading. London: ETI.

Ethical Trading Initiative (ETI) (2001) Annual Review 2000/2001 Learning our trade. London: ETI.

Fearne, A. and Hughes, D. (1998) *Success Factors in the Fresh Produce Supply Chain: Some Examples from the UK. Executive Summary*. London: Wye College.

Fine, B. (1994) Towards a political economy of food. *Review of International Political Economy*, 1, pp. 519–45.

Fine, B. and Leopold, E. (1993) *The World of Consumption*. London: Routledge.

Friedmann, H. (1993) The political economy of food: a global crisis. *New Left Review*, 197, pp. 29–57.

Gabriel, Y. and Lang, T. (1995) *The Unmanageable Consumer*. London: Sage.

Gereffi, G. (1995) Global production systems and third world development. In Stallings, B. (ed.), *Global Change, Regional Response: The New International Context of Development*. Cambridge: Cambridge University Press, pp. 100–42.

Glennie, P. and Thrift, N. (1996) Consumption, shopping and gender. In Wrigley, N. and Lowe, M. (eds), *Retailing, Consumption and Capital*. Harlow: Longman, pp. 221–38.

Goodman, D. and Watts, M. (eds) (1997) *Globalizing Food: Agrarian Questions and Global Restructuring*. London: Routledge.

Hartwick, E. (1998) Geographies of consumption: a commodity chain approach. *Environment and Planning D: Society and Space*, 16, pp. 423–37.

Hartwick, E. (2000) Towards a geographical politics of consumption. *Environment and Planning A*, 32, pp. 1177–92.

Hughes, A. (2000) Retailers, knowledges and changing commodity networks: the case of the cut flower trade. *Geoforum*, 31, pp. 175–90.

Ilbery, B. and Kneafsey, M. (1999) Niche markets and regional speciality food products in Europe: towards a research agenda. *Environment and Planning A*, 31, pp. 2207–22.

Ilbery, B., Holloway, H. and Arber, R. (1999) The geography of organic farming in England and Wales in the 1990s. *Tijdschrift voor Economische en Sociale Geografie*, 90, pp. 285–95.

Jackson, P. (1999) Commodity cultures: the traffic in things. *Transactions of the Institute of British Geographers*, 24, pp. 95–108.

Jaffee, S. (1995) The many faces of success: the development of Kenyan horticultural exports. In Jaffee, S. and Morton, J. (eds), *Marketing Africa's High-Value Foods: Comparative Experiences of an Emergent Private Sector*. Washington, DC: World Bank, pp. 319–74.

Jaffee, S. and Morton, J. (eds) (1995) *Marketing Africa's High-Value Foods: Comparative Experiences of an Emergent Private Sector*. Washington, DC: World Bank.

Johns, R. and Vural, L. (2000) Class, geography and the consumerist turn: UNITE and the Stop Sweatshops Campaign. *Environment and Planning A*, 32, pp. 1193–213.

Leslie, D. and Reimer, S. (1999) Spatialising commodity chains. *Progress in Human Geography*, 23, pp. 401–20.

Lockie, S. and Kitto, S. (2000) Beyond the farm gate: production–consumption networks and agri-food research. *Sociologia Ruralis*, 40, pp. 3–19.

Lowe, M. and Wrigley, N. (1996) Towards the new retail geography. In Wrigley, N. and Lowe, M. (eds), *Retailing, Consumption and Capital*. London: Longman, pp. 3–30.

Marsden, T., Flynn, A. and Harrison, M. (2000) *Consuming Interests: The Social Provision of Foods*. London: UCL Press.

Miele, M. (2001) Changing passions for food in Europe. In Buller, H. and Hoggart, K. (eds), *Agricultural Transformation, Food and Environment. Perspectives on European Rural Policy and Planning, Volume 1*. Aldershot: Ashgate, pp. 29–50.

Murdoch, J. (2000) Networks – a new paradigm of rural development? *Journal of Rural Studies*, 16, pp. 407–19.

O'Neill, P. and Whatmore, S. (2000) The business of place: networks of property, partnership and produce. *Geoforum*, 31, pp. 121–36.

Stevens, C. and Kennan, J. (2000) Will Africa's participation in horticulture chains survive liberalisation? Working Paper 106, Brighton: Institute of Development Studies.

Warde, A. (1997) *Consumption, Food and Taste*. London: Sage.

Wells, P. and Jetter, M. (1991) *Best Buys to Help the Third World*. London: Victor Gollancz.

Whatmore, S. and Thorne, L. (1997) Nourishing networks: alternative geographies of food. In Goodman, D. and Watts, M. (eds), *Globalising Food: Agrarian Questions and Global Restructuring*. London: Routledge, pp. 287–304.

2 Are hogs like chickens?

Enclosure and mechanization in two 'white meat' filières

Michael J. Watts

What happens when mechanization encounters organic substance?

(Giedion, 1948, 6)

In the zoo, as Berger famously put it, animals constitute a living monument to their own demise: 'everywhere animals disappear' (1980, 24). A central paradox resides in our experience of the zoo and its incarcerated inhabitants: the zoo is a source of appeal and considerable popular attraction and yet 'the view is always wrong...like an image out of focus' (Berger, 1980, 21). A double alienation lies at the heart of the zoo's paradoxical status. On the one hand it is a space of confinement and a site of enforced marginalization like the penitentiary or the concentration camp. On the other it cannot subvert the awful reality that viewed from whatever vantage point, animals are 'rendered absolutely marginal' (ibid., 22). The zoo recapitulates relations between humans and animals, between Nature and Modernity. It demonstrates a basic ecological fact of loss and exclusion – the disappearance and extinction of animals – through an act of incarceration.

Berger (1980) sought to connect the zoo as a monument to the loss of a specific social crisis: to the disposal, or perhaps more appropriately the enclosure, of the peasantry. In quick succession, the marginalization of the animal is recapitulated by the marginalization of a class for whom familiarity with and a wise understanding of animals is a distinguishing trait. For Hobsbawm (1994, 289), the second half of the twentieth century contained 'the most dramatic and far-reaching social change. ...which cut us off for ever from the world of the past, [namely] the death of the peasantry'. Whether the peasantry has died in such an irreversible way is a contestable point. Nevertheless, if the world-renowned San Diego zoo is a memorial to wildlife extinction and animal loss, so the monument to gradual disappearance of the European peasantry is the folk museum replete with artefacts and arcana of rural life. The zoo is a locus of pain and loss (Malamud, 1995, 179); the same might be said of the folk museum. Each represents an historical loss that is 'irredeemable for the culture of capitalism' (Berger, 1980, 26).

Berger's account offers a productive way of thinking about commodities and the commodification of nature. One might in fact construe the relation between animals and modernity as a gigantic act of *enclosure* containing a double-movement, at once cultural and political-economic. The first movement, of which the zoo is the exemplary modern instance, is to accomplish what Greenblatt (1991, 3) calls 'the assimilation of the other'. Animals within the zoo represent a cultural storehouse, part of the proliferation and circulation of representations which become 'a set of images and image-making devices that are accumulated, 'banked' as it were, in books, archives, collections, cultural storehouses, until such time as new representations are called upon to generate new representations' (ibid.).

This mimetic quality of capitalism suggests that representations of animals, ecosystems and nature are social relations of production. The representation is both the product of the social relations of capitalism and is a social relation itself 'linked to the group understandings, status hierarchies, resistances and conflicts that exist in other spheres of the culture in which it circulates' (Greenblatt, 1991, 6). The zoo is, then, product and producer (Malamud, 1995, 12). Its images and representations of animals achieve a reproductive power by speaking to our humanity, to the philanthropy of capitalists, to enhancing environmental sustainability, to perpetrating the distinctions between the wild chaos of Nature and the order of a rational capitalist world. The fact that zoos ultimately disappoint (Berger, 1980) is simply to assert that the hegemony of such representations is never fully secured. The zoo cannot address its contradictory embodiment of loss and captivity, the very antithesis of the fecundity and freedom which Nature purportedly signifies.

A second movement is political and economic. Berger (1980) gestures to it in his invocation of the relation between the animal world and the demise of the peasantry but Giedion (1948) poses the question directly in the quotation with which I began this chapter. The historical reference point here is the agrarian question of the mid- and late-nineteenth-century Europe. In Kautsky's (1899) still relevant account, the agrarian question encompassed accumulation (how were surpluses extracted from the peasant-dominated agricultural sector?), science (how were forces of production revolutionized?), and politics (what were the political implications for the German Social Democratic Workers Party of the disintegration and differentiation of the peasantry?). Kautsky (1899) emphasized that the 'disappearance' of the peasantry was a more complex and uneven process than orthodox Marxist predictions might suggest. Nevertheless, the historic demise of the peasantry was also a process by which agriculture was *industrialized*:

> bold prophets namely those chemists gifted with an imagination, already are dreaming of the day when bread will be made from stones and when all requirements of the human diet will be assembled in chemical factories.... But one thing is certain. Agricultural production has already been transformed into industrial production in a large number of fields. ... This does not mean that...one can reasonably speak of the demise of agriculture... but [it] is now caught up in the constant revolution which is the hallmark of the capitalist mode of production.
>
> (Kautsky, 1899, 297)

The second movement thus views the enclosure in class and industrial terms – what Giedion (1948) meant by mechanization. Peasant agriculture was differentiated from within and yet was simultaneously subject to twin processes of *appropriation* and *substitution* (Goodman *et al.*, 1987). The former spoke to a process by which more of the production process (including inputs) was provided by off-farm industry, while the latter identified an increasing trend to produce food and fibres in factories as a fully industrial process comparable to non-agricultural forms of manufacture. On one side stood drip irrigation and genetically modified crops, on the other the feedlot and the industrial manufacture of artificial sugars. The erasure of the peasantry/family farm as a social class is homologous to the industrialization and mechanization of animals within the food provisioning system. What Kautsky saw in its infancy was a process by which science was harnessed to the problem of the industrialization of livestock and crop production. Genetics laid the groundwork for the application of new technology to the plant and the animal *directly*. One might say that both the alienated, lethargic elephant depicted in Berger's (1980) description of the zoo and the genetically modified pig are monuments to the historic loss invoked by contemporary capitalism. One embodies the ecological crisis of modernity and memorializes the loss of biodiversity; the other captures an industrial-capitalist reconstitution of Nature and signals the extinction of the independent family farm.

The zoo and the mechanization of plants and animals can both be understood as forms of enclosure necessitating confinement, incarceration, discipline and subjection. Both necessitate an assimilation of the other, and the reproduction and circulation of mimetic capital through images, representations and fetishisms. For Berger (1980) the lion bred in captivity signifies the wave of modernity crashing onto the beleaguered peasant; for Gideon (1948) the genetically engineered broiler stands for the death of a Jeffersonian vision. Both the lion and the chicken are in their own ways cyborgs (Haraway, 1991).

This chapter explores two contemporary agro-food commodity chains – that of the US chicken (broiler) and the hog – in light of processes of enclosure and mechanization, and considers how each ramifies through two differing commodity systems (*filières*). My focus primarily is upon political-economic rather than cultural-symbolic arenas with a concrete interest in contrasting commodity dynamics. From a methodological vantage point, one of the greatest difficulties for analysis is posed by the synchronic and static qualities of the commodity chain. The chapter thus seeks to address questions about process, trajectory and historical dynamics by comparing two rather different commodities. In doing so I identify six 'moments' of the agro-food commodity chain: *commodity profusion, commodity subsumption, commodity integration, commodity qualification, commodity concentration*, and *commodity trajectory*. At the heart of my analysis is the dual question of commodity convergence and replication: are hogs really no different than chickens and is the history of the industrial hog a tardy replication of the history of the industrial chicken? That is, does the biology of the commodity make a damn bit of difference?

A note on enclosure

In the English language the word enclosure has a strong historical referent, namely the English Parliamentary enclosures of the eighteenth and nineteenth centuries. Until the eighteenth century English villages were distinguished by the fact that land for cultivation and grazing was shared in some way. To read John Clare's early nineteenth century poetry is to experience a natural world known to him in virtue of its openness. Sharing, common access and possession without ownership of land provided a social integument to village life in Helpston:

> Love hearken the skylarks
> Right up in the sky
> The suns on the hedges
> The bushes are dry
> The slippers unsullied
> May wander abroad
> Grass up to the ancles
> Is dry as the road
>
> There's the path if you chuse it
> That wanders between
> The wheat in the ear
> And the blossoming bean
> Where the wheat tyed across
> By some mischevous clown
> Made you laugh though you tumbled
> And stained your new gown
> (Clare, 1967, 39)

Commoning economies involved common *right*: access to pasture, waste and other village lands through customary (and *de facto* legal) forms of usufructory right. These rights provided the English commoner some measure of independence from both the labour market (having to work as labourers) and the cash economy (having to purchase the means of subsistence). Much of England was still 'open' in 1700, and the commons were intact; by 1840 most land was enclosed and the commons virtually obsolete. In Clare's Helpston, fields of wheat and beans were now fenced and railed, while in more pastoral areas rights to graze were abrogated. Enclosure produced a form of confinement:

> These paths are stopt – the rude philistines thrall
> Is laid upon them and destroyed them all
> Each little tyrant with his little sign
> Shows where man claims earth no more divine
> But paths to freedom and to childhood dear
> Aboard sticks up to the notice 'no road here'
> And on the tree with ivy overhung

The hated sign by vulgar taste is hung
As tho the very birds should learn to know
When they go there they must no further go
This with the poor scared freedom bade goodbye
And much they feel it in the smothered sign
And the birds and trees and flowers without a name
All signed when lawless laws enclosure came

(Clare, 1982, 415)

The controversial enclosure acts which gained momentum from the 1740s onward had devastating effects on commoners. Nature was transformed biophysically as well as socially through the loss of particular customs, rights and forms of livelihood. Boundaries and borders restricted movement and altered the experience of the landscape. Juridical and economic derelictions hastened the end of a certain sort of moral economy, of a particular sociability and of a sense of community. None of this is to wax nostalgic or to mythologize relations of exploitation and subordination. It is rather to seek to understand a process of simultaneous natural, physical and socio-cultural transformation.

Enclosure is both a powerful historical fact and a carrier of a dense cluster of meanings which refer in part to a spatial process. Place and territory is reconstituted through boundaries and property lines. But enclosure also spoke to the loss of rights, obligations and responsibilities. Enclosure involved a loss of the 'open' commons; it simultaneously wrought dispossession, incarceration, privatization and social transformation. Understood in this way, enclosure comes to speak for the social, the spatial and the natural all at once. It also is a contradictory process: as it closes it also opens. It both excludes and confines; not everyone experiences dislocation in the same way. The enclosure movement had its origins in a particular sort of modernity, and insofar as it is geographical, makes space and space-making central to the idea of the modern. My exploration of the relations between enclosure and contemporary food-commodity chains sees enclosure operating in four primary ways: through production relations at the level of the grower (*integration*), at the level of the animal and bird growth and maturation (*confinement*), at the level of breeding and organic substance (*genetic enclosure*), and through the market (*industrial concentration*). These pathways represent distinctive geographical sites of enclosure – spaces of differing sorts of capture.

Feathered friends

The chicken industry has been one of the most dynamic sectors in the post-war US economy. In terms of biological productivity, mechanization and industrialization the chicken has no peer in contemporary food provisioning (Boyd, 1998, 2001). Since the early 1930s the broiler industry has been transformed from a backyard industry of 34 million birds valued at $18 million into a tightly integrated, highly specialized and corporate-dominated transnational agribusiness producing 8.262 billion broilers valued in 2000 at $13.953 billion (NASS 2001).[1]

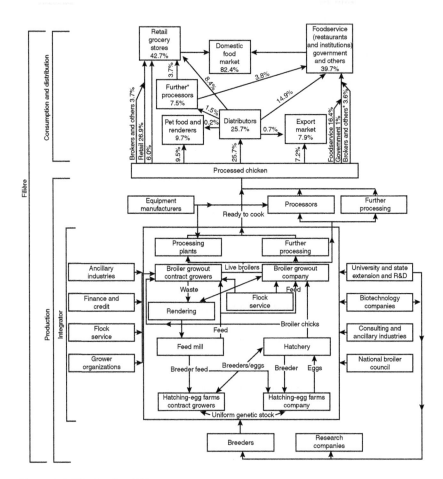

Figure 2.1 The broiler filière.

Source: Boyd and Watts, 1997, 205.

The broiler filière is as a consequence a complex institutional nexus of actors (Figure 2.1). US per capita chicken consumption currently stands at 78.8 lbs (by comparison the figures for beef and hogs are 69.4 and 52.5 lbs respectively), roughly *triple* of the level in 1960! In 1991, chicken consumption per capita exceeded beef, for the first time, in a country which has something of an obsession with red meat.

The fact that each American man, woman and child currently consumes roughly 1.5 pounds of chicken weekly reflects a complex vectoring of social forces in post-war America. First, a change in taste has been driven by a heightened sensitivity to health matters and especially heart-related illnesses associated with red meat consumption. Second, increased consumption has been stimulated by the fantastically low cost of chicken meat (driven by massive productivity increases) which has in real terms *fallen* since the 1930s. Finally, the industry has experienced *commodity profusion*. Chicken is processed and consumed in a huge variety of forms which did not exist twenty years ago, and are now delivered to consumers by the gargantuan fast-food industry (Schlosser, 2001). In 1963 whole birds represented 84.4 per cent of sales; by 1997 the corresponding figure was 13.1 per cent (Ollinger *et al.*, 2000).

Standing at the centre of the commodity chain (Figure 2.2) is the 'integrator', the large company which has integrated the specialized stages (breeding, feedstuffs,

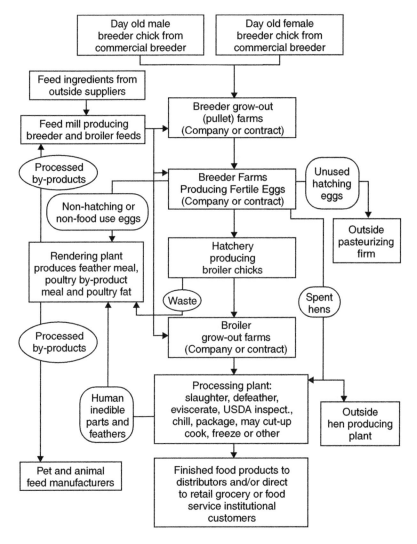

Figure 2.2 Vertically integrated broiler filière.

Source: Adapted from Roenik, 1991.

'growout', slaughter, processing) of the broiler complex (Tweeten and Flora, 2001). The integrators – the outcome of *commodity integration* – dominate the entire industry. In 2000 the top three integrators accounted for 40.5 per cent of production and 46 per cent of broiler slaughter and processing. Corporations such as Tysons, Goldkist, Perdue and Foster Farms emerged following the 'backward feeder integration model' pioneered by Jesse Jewel in Georgia during the inter-war period. Over time mergers and consolidation have generated a striking *commodity concentration*. In 1955 the top four firms accounted for 12 per cent of broiler production; by 2000 the figure was 47.5 per cent. Within the 1990s alone the three-firm concentration ratio increased from 35 per cent to over 40 per cent. The top 23 (of 54) integrators accounted for 87 per cent of broiler production by 2000 (Watt Poultry, 2001). The dominant company in the industry, Tyson Foods, accounts for 41.6 per cent of total slaughter and controls 23.5 per cent of the US market. The corporation's stated aim is to 'control the centre of the plate for the American people' by 'segmentation, concentration and domination' (Don Tyson, CEO, cited in Koonce and Thomas, 1989). Two-thirds of Tyson's broiler output is destined for the fast food industry.

Broilers are overwhelmingly produced by family farmers ('growers') but this turns out to be a deceptive description. They are in fact raised from day-old chicks to 45-day (4.8 lbs live weight) broilers by farmers under contract to multi-billion dollar transnational integrators who own the chicks and feed. Non-unionized growers must borrow heavily in order to build the infrastructure necessary to meet rigid contractual requirements intended to ensure 'quality'. Conventional contract terms are such that integrators provide growers with chick or poult hatchlings and feed from integrator-owned hatcheries and feed mills, and veterinary services, medication, litter and field supervisors. Conversely, contract growers provide housing equipment, labour, water and fuel (Figure 2.3). The average broiler grower is a 48 year old male[2] who owns 103 acres of land, 3 poultry houses and raises 240,000 birds under contract through six flocks per year; he owes over half of the value of the farm to the bank and works more than 2,631 hours per year. A grower's net farm income from poultry is about $15,000 annually, less than half that of comparable non-poultry farms (Perry *et al.*, 1999; www.web-span.com) (see Figure 2.4). Since 1995 returns to growers have been declining as a result of increasing feed costs, flat demand and a reduction in export markets (Perry *et al.*, 1999).

Growers are themselves differentiated and their composition and scale of production has changed dramatically in the last half century. Farm size has increased as the number of broiler farms have fallen. In 1959 there were 42,185 broiler farms with only 5 per cent raising 100,000 birds or more. By 1987 over half of the 27,645 growers were raising more than 100,000 birds. More than a quarter of total US broiler sales was raised by 4 per cent of growers. By 1999, small broiler farms (less than $100,000 gross sales) constituted 54 per cent of farms and only 12 per cent of output by value; farms with sales between $100,000 and $999,999 accounted for 75 per cent of output. The largest 3 per cent of producers accounted for fully one-third of total broiler production (Perry *et al.*, 1999, 7).

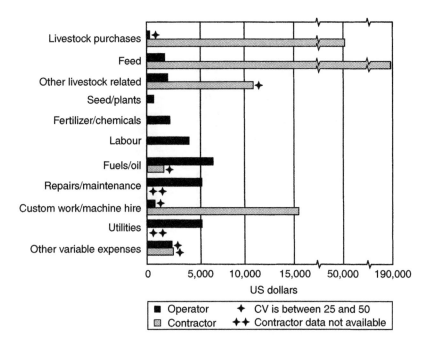

Figure 2.3 Distribution of variable expenses between operator and contractor on broiler farms, 1995.

Sources: Farm Costs and Returns Survey and ERS/USDA, 2000.

By 1994 contract production represented 85 per cent of chicken output. Only 14 per cent was accounted for by the companies themselves in integrated operations. Contracts vary substantially but integrators typically agree to pay a pre-established base fee per pound of live broiler plus a bonus or penalty for performance relative to other growers (Knoeber, 1989). The performance bonus is calculated on the difference between actual grower settlement costs and the average settlement costs of other growers harvesting flocks at that time. The total grower payment is determined mainly by the feed-conversion ratio and disease/environmental bird mortalities (Ollinger *et al.*, 2000, 12). Integrators establish contracts with numerous growers usually located within 20 miles of an integrator-owned slaughter/processor/packing plant. They control supply by regulating the numbers of chicks placed in 'growout' while controlling the quality and physio-morphological character of birds appropriate for an increasingly mechanized and routinized slaughter and processing industry. Contract growers thus are not independent farmers at all. They are little more than 'propertied labourers': employees of corporate producers who also dominate the processing industry.

The slaughter/processing industry has undergone restructuring in tandem with integration and the productivity revolution. By 1968 the basic automated slaughtering process had been established in which live birds are stunned or

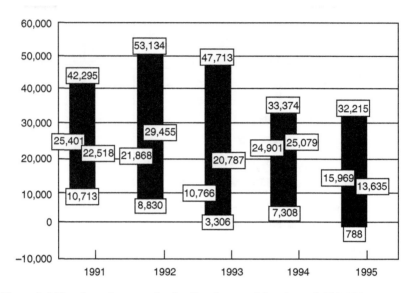

Figure 2.4 Net farm income for broiler farms with sales of $50,000 or more, 1991–95.

Sources: US Department of Agriculture's 1995 Agricultural Resource Management Study (previously known as the Farm Costs and Returns Survey) and ERS/USDA, 1999.

Note
Graph shows the top 75 per cent level and bottom 25 per cent level on each bar, as well as mean income to the left and median income to the right of the bar.

gassed, cleaned, iced, cut up, refrigerated and cut up, boned or processed. Automation has changed few aspects of bird input but outputs have altered markedly. Whereas in 1963 a plant sold a whole bird to a retailer, whole birds represented only 13 per cent of slaughter by the 1990s (Ollinger *et al.*, 2000, 13). Economies of scale and increase in plant size coincided with commodity concentration. As poultry plant size grew, the number of growers dropped by 33 per cent between 1972 and 1992 and their average size tripled.

Like broiler production, the processing sector is highly regionalized. Two-thirds of chicken slaughter products are produced in the southeast (up from 55 per cent in 1963) (see Figure 2.5). In all locations, working in the poultry processing industry is one of the most underpaid and dangerous in the country. Vietnamese, Laotian and Hispanic immigrants now represent a substantial proportion of almost wholly non-unionized workers, especially in the US south. In a Federal indictment submitted in Chatanooga, Tennessee in December 2000, Tyson's was accused of smuggling illegal immigrant workers across the Mexican border to work in processing plants (Barboza, 2002).

Despite the rise of commercial hatcheries early in the twentieth century, the chicken industry remained a sideline business run by farmer's wives until the

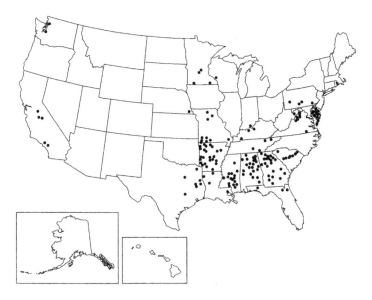

Figure 2.5 Locations of major broiler processing/further processing plants.
Source: Poultry Digest; internet homepage (http://www.wattnet.com).

1920s. Since commercial sales began, the industry has been transformed both by feed companies which promoted integration and the careful genetic control and reproduction of bird flocks, and by scientific developments often supported by the federal government. Until the Second World War, the chicken industry primarily was located in the Delmarva Peninsula near Washington, DC, however during the 1940s and 1950s the geographical expression of the industry's *commodity trajectory* changed. The contemporary heart of the US chicken industries is in the former slave holding and cotton growing South: Georgia accounted for 15.2 per cent of all broilers in 2000, and four southern states (Alabama, Mississippi, Georgia and Arkansas) accounted for 50.8 per cent of the US total. Industrial structures also were transformed with the emergence of large integrated broiler complexes at the forefront of what is now called flexible or just-in-time organization.

The post-war history of the national and global chicken or 'broiler' industry is one of exceptional dynamism and growth, characterized by *commodity subsumption*. Advances in breeding, disease control, nutrition, housing and processing have resulted in some of the lowest production costs in the world for US integrators. Since 1940, the industry's feed conversion rate (pounds per feed per pound of live broiler) has declined from 4.30 to 1.75. The average broiler live weight has increased from 2.89lbs in 1934 to 4.86lbs in 1998 and the time required for a bird to reach market weight has plummeted from over 95 to days 45. Avian science has enabled birds to add weight at mind-boggling rates of almost five pounds in as many weeks. In the last 50 years, average live bird weight has doubled while the labour input in broiler meat production declined from 5.1 hours per 100 lbs in 1945 to less than eight minutes in 2000! (USITC 1992).

The broiler is a product of the massive research and development campaign increasingly driven by genetic engineering. Genetic modification, breeding, high nutritional growth regimes and disease regimens have produced a steroidal chicken designed for both the machine and the market. In order to maximize the production of breast meat and to facilitate mechanized slaughter and cutting, radical changes in the skeletal form of the chicken have been required. The broiler has thus become a grotesque Frankenstinian product of biotechnical mechanization (the nutrition–feed–health–confinement complex), consumer preference, and the rigors of competitive capitalism.

Two forces were central to radical increases in biological and labour productivity, and to the reconstitution of the broiler itself. First, the move to year-round spatial immurement – *commodity confinement* – was made possible by the discovery of vitamin D in 1926, and facilitated the shift to industrial broiler production. Second, post-war productivity increases rested heavily upon activities of the US federal government during the depression years. Although chicken breeding experiments had been established by the US Department of Agriculture (USDA) as early as 1912, it was not until the establishment of the National Poultry Improvement Plan (NPIP) in 1933 that the government emerged as a major force in developing new disease control, breeding and husbandry techniques. A gradual reduction in mortality rates permitted the move to more intensive confinement operations. Even more decisive was a revolution in primary breeding (Boyd, 2001; Bugos, 1992). Until the 1940s, farmers had relied on pure breeds developed for egg production, with little concern for meat qualities. Increased demand for chicken meat during the war led breeders to focus attention on the development of specialized breeds for meat production. Employing principles pioneered in the hybridization of corn and other crops, primary breeders developed standard pedigrees of male and female lines that combined to produce a superior bird. By the 1950s such cross-breeds provided the genetic basis of the modern broiler industry, and breeders sought to fine-tune pedigrees to meet ever more exacting demands for genetic uniformity and quality assurance – *commodity qualification*.

Complementing dramatic advances in breeding were substantial public investments in nutrition, disease control and confinement technologies during the immediate post-war period, much of which was sponsored by the land grant universities and the federal government (Boyd, 2001). The development of high-performance rations and the use of vitamin B-12 and antibiotics in feed dramatically reduced mortality and increased feed conversion efficiency. By the 1950s a combination of confinement, nutrition and growth research and genetic improvement had produced a chicken 'which was *made to order to meet the needs of the meat industry*' (Warren, 1958, 16, emphasis added). By the 1980s new biotechnologies entered the broiler industry, led by life sciences companies intent upon genetically engineering breeds keyed to the sale of proprietary health products. With the famous Chakrabarty ruling in 1980, substantial interest was aroused in the possibility of transgenic chickens subject to patent protection. Completion of the full genetic map of the chicken would 'allow selective improvement to proceed on the basis of genotype rather than phenotype, representing a very significant expansion in "breeding power"'(Boyd, 1998, 34).

The chicken-genome project marks a century-long process of commodity enclosure. First, the notion of confinement speaks both to the shift from open range to broiler houses and a process of industrial integration within a centralized complex, which has as its counterpoint the enclosure of the chicken contractor (Boyd and Watts, 1996; Watts, 1991). Second, the emergence of the 'designer chicken' establishes the extent to which nutritional and genetic sciences have produced a cyborg broiler to fit the needs of the industry. Third, the chicken itself is transformed into a site of accumulation, reflected in its curious physiognomy and anatomy ('all breast and no wings'). The 'working body' (Harvey, 1998) of the chicken has not simply been 'Taylorised' but actually constructed physically to meet the needs of the industrial labour process. The poultry industry in this sense combines the worst of productive consumption of the human body (the appalling working conditions and health deficits associated with working on the line) with the most grotesque forms of reconstituted Nature. Nineteenth century work conditions meet up with twenty-first century science.

Finally, biological and environmental resistances to commodification (see Polanyi, 1944) are nowhere clearer than in the broiler industry. As demonstrated by the Hong Kong-derived avian flu of 1998, the dangers of disease are driven in part by the susceptibility of fragile and vulnerable chicken to new pathogens which are quickly able to destroy flocks. Efforts to increase breast meat yields have created a high propensity for muscular and skeletal problems, metabolic disease, immuno-deficiency and male infertility (Boyd, 1998, 15), generating high bird mortalities. At a social level, the health problems of the industry are legion. In the United States where the salmonella problem and chicken are virtually synonymous, antibiotic use as feed additives has remained largely unregulated. A popular concern with labelling and a return to free range chicken sharpens the contradictions of the industrial and productivist model of agriculture (see also Guthman, Chapter 12, this volume).

Confinement, battery-hens and the suffering imposed by mass production have raised considerable ethical questions.[3] Peter Singer, who once sat in a cage in the middle of Melbourne to publicize the plight of battery hens, specifically argues that the 'simple' chicken deserves to be protected from unnecessary pain (*New Yorker*, 6 September 1999). Animal welfare concerns are no longer local or parochial matters; indeed they have reached the bargaining table of the World Trade Organization. The European Union sought to pressure trade partners to meet its own animal welfare standards because of a concern to limit animal rights considerations against the 'competitive position of EU producers in a market liberalized under WTO agreements' (Fisher, 1999, 13). Nonetheless, on 15 June 1999, the EU Agriculture Council finally decided to prohibit the use of battery cages by 2012. Qualification of the commodity chain is here driven by ethical considerations.

And corn-fed porkers

At first glance, the US pig filière appears an unlikely candidate for a case of commodity convergence with lowly chicken. The hog has a much deeper history of commercialization and stood at the core of the Midwestern US meat processing industry in the nineteenth century. By the 1830s Cincinnati had emerged as 'Porkopolis', a major hub of the meat processing revolution in which the pig was disassembled in increasingly mechanized plants. Long before Chicago became the industrial centre, a commercially oriented pig–corn complex had developed in Ohio, Illinois and Indiana sustained by family farms and linked via a complex network of brokers and merchants to processors. In the nineteenth century the pig stood at the centre of the processing revolution (Pudup, 1983).

Yet by the 1990s the pork industry in the US was moving away from a traditional structure of hundreds of thousands of largely mid-Western family farms selling at local spot/terminal markets to a 'more concentrated supply chain model' (Drabenstott, 2000; Page, 1997). The revolution resembled the restructuring of the broiler industry 40 years earlier, as contraction, integration and concentration rapidly developed. Vertical coordination and integration associated with new technology, geographical shifts in production and growth in operations have occurred alongside substantial increases in productivity and efficiency (Figure 2.6).

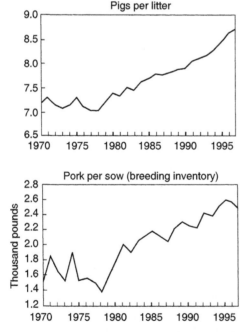

Figure 2.6 Efficiency gains in hog production.

Source: Economic Research Service, US Department of Agriculture.

Contracting of pig production by large packers is now a major feature of the commodity chain. In 1960 less than 2 per cent of hogs were produced under production contract or in vertically integrated firms; by 1999 the figure was almost 60 per cent (Martinez, 1999). In a manner unthinkable even 20 years ago, the pork commodity-chain is now dominated by debates about vertical coordination and the 'industrialization of agriculture' (Martinez *et al.*, 1997). The rise of the industrial chicken has seen the rise of its animal counterpart, the industrial hog (genetically modified, steroidally enhanced, confined and incarcerated), several decades after the broiler-cyborg.

At the farm level, hog production generated $13.2 billion in farm revenues in 1998. If processing is added, pork is a $28 billion industry employing 600,000 people. The historic heart of production resided in the Corn Belt and in particular Iowa, where family farms raised pigs from birth to market. By the 1970s, specialization had segmented the process into three discrete stages. The first segment or site is breeding gestation and farrowing. After weaning pigs are moved to a nursery site where they receive special nutrition to reach 12–16lbs. Once they reach 8–10 weeks and 40–60lbs, pigs are transported to a finishing facility where they are fed in confinement to market weight (ca. 250lbs) between the ages of 150–210 days. In the last three decades these sites have been coordinated and integrated in radically new ways by 'producer–integrators' (Figure 2.7) giving birth to a hog-complex structurally similar to the southern broiler-complex (Boyd, 2001). In the same way that broiler integrators such as Tysons stand at the centre of the filière, the firms of Premium Standard farms, Seabord and Smithfield have come to dominate the pork commodity chain.

Specialization and segmentation in the pig filière also has been driven by the genetics–nutrition–health–confinement revolution – what I referred to

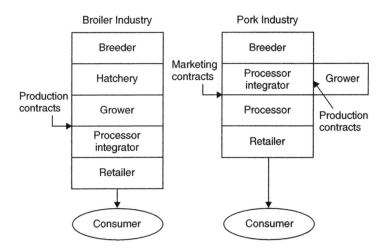

Figure 2.7 Production and marketing contracts in the pork and broiler industries.
Source: Economic Research Service, US Department of Agriculture, 1999.

previously as *commodity subsumption*. In the same way that the organizational and biological rearrangement of the broiler has enabled the more rapid and lower cost production of breast meat, so too a new generation of pork genetics produces leaner pork more cheaply on carcasses that have been literally constructed to meet the mechanized needs of the processing industry.[4] In the last 20 years, fat in pork chops has been decreased by more than a third. At the same time the industry has witnessed a striking productivity revolution. Since 1970 pigs per litter has increased from 7 to 8.8 in 2000 and pork per sow from 1,500 to 2,600lbs. Feed efficiency has also improved markedly. As a consequence, between 1955 and 1998 the retail price of pork fell from $2.60 per lb. to less than $1.30.

With powerful scale economies at work in the filière, pig production and processing has been dramatically transformed (Horwitz, 1998). Between 1950 and 1998 the number of pig farms fell from 2.1 million to 139,000 and concentration proceeded apace (Figure 2.8). The largest hog farms (50,000 hogs a year or more) accounted for 37 per cent of total US output in 1998 (up from 7 per cent in 1988). The proportion of all hogs produced on farms of 5,000 hogs or more doubled between 1995 and 2000. In new grower states such as North Carolina, 98 per cent of all hogs resided on these large farms (compared to 63 per cent in Iowa, the leading hog-producing state in the old Corn Belt) (NPPC, 1996; USDA, 2001). The average size of the production operation also has increased. In 1949 the modal hog farm in Iowa and North Carolina sold eighty-three and eleven hogs respectively; in 1992 the figure was 787 and 2,686.

Hog integrators have driven the industrialization and mechanization process in two discrete ways. First, marketing and production contracts with growers' 'finishers' are almost identical to broilers (Table 2.1). Integrators provide management services, feeder pigs, health and other services and key inputs, while growers supply labour and facilities. In return growers receive a fixed payment

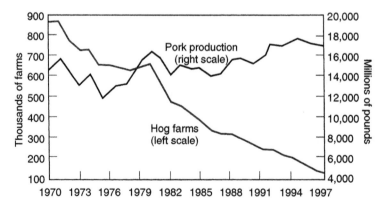

Figure 2.8 Pork production and the number of hog farms in the United States.
Source: Economic Research Service, US Department of Agriculture.

Table 2.1 Hogs farrowed and finished by producers using production contracts, 1997[a]

Size class (1,000 head)	Total farrowing (%)	Total finishing (%)	Contract farrowed (%)	Contract finished (%)
1–49	10	14	1	8
50–499	8	9	4	7
500+	22	22	11	16
Total	40	44	17	30

Source: Lawrence, J., Grimes, G. and Hayenga, M. *Production and Marketing Characteristics zof U.S. Pork Producers, 1997–98*, Staff Paper 311, Iowa State University, Department of Economics, December 1998.

Notes
a Total farrowing and total finishing represent the percentage of total production from operations of producers who raised their own pigs plus their contracted operations (i.e. as contractors, they had other producers raise their pigs under production contracts). Contract farrowed and contract finished represent the contract production of these producers as a percentage of all production.

adjusted for 'production efficiency'. Marketing contracts typically specify that growers deliver a number of hogs to a particular location at a particular time subject to quality premiums. Developed in the hog sector by Wendell Murphy[5] in North Carolina in the 1960s, contracts began among the large operations of the so-called East Coast Region and have recently moved to the wide west and Great Plains. One third of east coast growers had entered contracts before 1985; the comparable figure for the mid-West was less than 10 per cent (Kleibenstein and Lawrence, 1995). In the early 1990s the number of hogs grown under contract was increasing by 30 per cent per annum.

A second driver of industrialization and mechanization has been vertical integration in which large breeder–grower–processors control all stages of the process. While only 4 per cent of hogs are currently produced in this fashion (there was virtually no integration in 1980), it is growing rapidly, notably in the frontier western states like Colorado and Oklahoma. Overall, the top 10 integrators currently account for over 35 per cent of the nation's inventory (up from 16 per cent in 1995).

Concentration in hog packing and processing is equally striking. In the early 1900s, most hogs were slaughtered by packers who purchased pigs through commission firms located in public terminal-markets adjacent to their plants in Chicago and Indianapolis. This system, now extinct, was replaced by spot-markets – although by the year 2000 spot markets themselves accounted for only 26 per cent of transactions (down from 43 per cent in 1997). Nearly 75 per cent of all hogs currently are sold under marketing contract (typically ledger contracts) between a producer (who owns hogs) and a packer. The specific price, however, is determined by a new carcass merit system rather than live weight. Using electronic grading devices such as optical probing, ultrasound and magnetic resonance imaging, discounts and premiums are awarded to growers in relation to specific qualities (backfat, muscling, colour, pH, tenderness). Nonetheless the size of the supplier and the packer determines the forms of market or non-market transaction. Smaller volume producers typically use the spot market while some of the largest packers are increasingly producing their own hogs in-house; by the year 2000 the number of hogs owned and slaughtered by packers had reached 24 per cent. Like the broiler sector, hog packers and processors have aggressively marketed a range of new pork products tightly linked to the fast-food industry (Schlosser, 2001).

The integration and specialization of the commodity chain has resulted in a restructuring of the geography of production. This new geography has been associated with extraordinary inputs from the corporate welfare state and local government (tax breaks, subsidized inputs, local provision of roads, airports).[6] While Iowa has retained its role as the largest grower of hogs in the nation, the relative significance of the Corn Belt has diminished in the last three decades (Thu and Durrenberger, 1998). In 1960 hog production was scattered across most states but nine Corn Belt states accounted for 65 per cent of output. By 2000, 46 per cent of hogs were produced in Iowa and North Carolina alone. The North Carolina industry is characterized by highly coordinated contractual arrangements and mega-sized operations. Mature and medium to small family farm producers in the mid-West, and in the Iowa heartland, began to lose their competitive edge in the 1970s. Vertical coordination and contracts *are* proliferating in the region but unevenly and with different configurations from the North Carolinian mega-operations. Iowa appears to be specializing in the grow–finishing phase of swine production (because of cheaper corn prices) while the fringe Corn Belt states to the southeast, south and southwest are increasing their share of earlier phases. Slaughter capacity also has shifted with production: Smithfield's Foods opened one of the world's largest packing plants in North Carolina.

Most recently pork production has shifted spatially once again, moving into the Texas and Oklahoma panhandle, southwestern Kansas and southern Utah.[7] Pork production in Oklahoma has grown by 900 per cent in a decade! Massive new operations and processing facilities opened in the 1990s, with Tyson's adopting a contractor model and Seabord a more vertically integrated approach (Mayda, 2001). Much new expansion is undertaken by the very largest operations (Smithfield, IBP), employing contract production and often large integrated production facilities. For 3,000 plus sow operations, the US West has some of the lowest production costs, the weakest environmental regulations and the geographical proximity to west and increasing Asian export markets.

Dynamism and change within the pork filière – the fact that a given quantity of pork meat can now be produced with fewer hogs, less labour and less feed – primarily has been based upon declining or static domestic per capita consumption between 1970 and 2000 (roughly 49 pounds per capita). Total output has also only increased by 12 per cent over the same period. Since 1990, pork production in the United States has grown by 2 per cent per annum, driven by a five-fold increase in exports especially to Japan, Canada and Mexico.[8] The hog filière is in this sense, *contra* the chicken, a case of revolutionary mechanization *without* radical output changes; what one might call an 'industrialization without growth' trajectory. Like chicken, however, the industrial hog has generated enormous health and environmental consequences (Thu and Durrenberger, 1998). The use of antibiotics in hogs is widespread and the controversy about sub-therapeutic use is longstanding (see Halverson, 2000). Toxic and health considerations surrounding so-called waste lagoons have made new hog operations the source of intense community, legal and regulatory struggles (www.hogwatch.org). North Carolina, for example, enacted a moratorium in 1997 on the expansion or start up of pork facilities with more than 250 hogs. The law also directed the state Environmental Protection Agency to more closely regulate odour emissions. As a result, 'firms in the pork industry are comparing regulatory climates across states lines and even county borders in search of places with fewer regulations' (Drabenstott, 2000, 91).

Comparative enclosures

The post-1945 recomposition of the broiler and hog filières is a striking case of commodity convergence – what Friedmann (1994) calls 'replication'. Chains have been industrialized along their length, and the genesis of the industrial pig and chicken complex stand as testaments to Giedion's (1948) question about the encounter of mechanization and organic substance. At the heart of each filière stands the integrator, in both cases hatched in the US south, and led by the aggressive activity of feeder, and subsequently processor, capital. In short measure, the chain underwent a rapid specialization and integration to produce a vertically coordinated, flexible organizational configuration. For both hogs and broilers the shell of the family farm is both preserved and transformed through the contract into a sort of worker or company employee in which the much-vaunted autonomy and independence of the family farm is lost.

The 'New South' – North Carolina, Arkansas, Georgia – figures centrally in the trajectory of the filières. This is a region with a long and oppressive history of share contracts, despotic labour relations and poverty. Broiler and hog complexes each have been established on the back of another commodity crisis: in the case of chickens the collapse of cotton, and for hogs the collapse of tobacco. The astonishing rise of the chicken in Georgia and of the hog in North Carolina thus represent *commodity displacement*.[9] As the filière deepens and matures, competitive pressures and internal resistances push the system into new and often more marginal and impoverished regions. In the case of chickens this has been the recent expansion into rural Mississippi, and for hogs into the Panhandle. The replication argument is granted even more plausibility by the recent integration of the two filières themselves, as large processors from the chicken sector (most notably Seaboard, Tyson's and ConAgra) inexorably moved into pork during the 1990s.[10] What is at stake now is a sort of *trans-commodity integration*.

The chapter has characterized the rise of the mechanized broiler and hog and the recomposition of their commodity chains through subsumption, integration, differentiation, qualification and specialization as a sort of multifaceted enclosure, rooted in processes of animal/bird confinement, grower contractualization, genetic enclosure and market concentration. However this picture of confluence and replication is at once too neat and insufficiently dynamic. Hogs have not duplicated chicken in a mimetic way. First, coordination and integration has developed further, in part because of the expansion of just-in-time systems (see Boyd and Watts, 1997). Spot markets still flourish for hogs, and the complexity and depth of contracts associated with chicken has no parallel in the pork sector. A great stumbling block has been the robustness and stability of extant production systems in the Corn Belt, *the* low cost zone for feed. Integration will depend on the extent to which the family farm sector and the historic farrow-to-finish systems in Iowa can be broken; and this is a political and economic question. Second, the broiler productivity revolution was predicated on a massive expansion of output and market deepening, whilst static domestic demand for hogs has placed outer limits on processes of internal restructuring. Differing chain dynamics also account for contrasting periodicities and cycles of overproduction within both chains.

Third, we must consider points of tension within filières. A coordination-integration tension lies along one axis. In the case of broilers, Tyson's outgrower contract model has no parallel in California, where integrated production systems have become dominant in part because the state had no family farm sector. Differing firm strategies in Oklahoma and elsewhere also indicate tensions between coordination and integration within the hog filière. Along another axis lie political resistances thrown up by the industrialization of the commodity chain. The chicken sector has been characterized by intense competition between regional producers over market share, specifically access of fresh (non frozen) chicken to large regional markets. Western US growers have become increasingly organized and militant. Conversely, resistances across the hog commodity chain are associated with community struggles over massive operations in new frontier areas, and with the devastating environmental consequences of a waste-intensive industry. Future transformations also will be dependent upon the extent to which hog producers can propel the robust and politically organized Corn Belt farmers into the integrator system.

Finally we must reflect upon the historical dynamics of commodity chains. In chicken production, the contract/organizational revolution and a shift from the mid-Atlantic states was almost complete by the 1950s (Figure 2.9). The hog followed three decades later. A lag in the recomposition and reorganization of the filière turns in part on biology and genetics. Not only can chicken genes be manipulated more easily but also the two animals have differing biological lags in production time[11] (see Figure 2.10). Further, integrators confronted different relations of production in 'new' and 'old' locations. Paths to integration and coordination had different rhythms, different speeds and periodicities and different internal dynamics. Restructuring of the southern chicken industry did create a form of path dependency for the hog that followed. To peer down on the hog and broiler industries from the vantage point of the twenty-first century is to recognize the vastly powerful forces of replication – enclosure and mechanization – at work across the modern US agro-food sector. At the same time, however, we must not lose sight of differences in biology, history and region that grant commodity chains their specific gravity, colour and particularity.

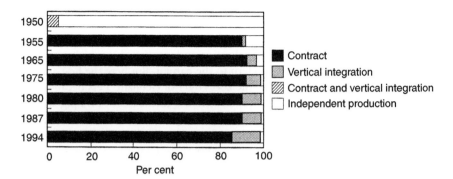

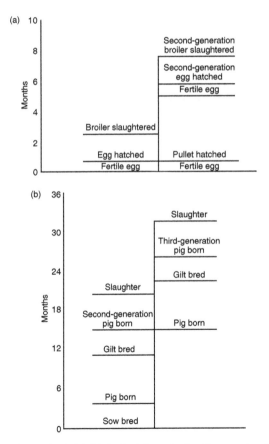

Figure 2.9 Proportion of broilers produced under contracts, vertical integration and independent production.

Source: Economic Research Service, US Department of Agriculture, 2000.

Figure 2.10 (a) Biological lags in broiler production. (b) Biological lags in hog production.

Source: Adapted from Roenik, 1991.

The contemporary US landscape of hogs and chicken returns us inevitably to Berger's (1980) sense of loss, marginalization and incarceration which somehow recapitulates the relation of the modern to the natural. But here the hog and the broiler are both more and less than the zoo's tiger or elephant. Less because they are less visible to the consumer (who has visited a chicken farm or a pig slaughter house?). And more because 'white meats' speak to more than salvage or preservation. Each assuredly is a marker of a certain sort of loss – particularly of the small, independent family farm. However these commodity chains are also *produced* (not just constructed or captured) materially, biologically and symbolically; they are the artefacts of modern capitalism as much as DVDs or automobiles. The fact that the US premedical curriculum of biochemistry, cell biology and genetics is based on only seven species (one of which is the chicken) is a measure of the distance that the chicken and the hog have travelled from the zoo, and from Berger's sense of exclusion.

Acknowledgements

This paper began long ago with some collaborative work on chickens with William Boyd of Stanford University Law School (Boyd and Watts, 1997). Brian Page and Leland Southern of USDA/ERS, Washington, DC supplied information on the US hog sector. Jacki Kohleriter provided research assistance.

Notes

1 The broiler is a young chicken raised for slaughter. Broilers represent two-thirds by value of the poultry industry, and 75 per cent by value of the chicken (broilers and egg production). My remarks are devoted to broilers and I shall exclude egg production which has many similar attributes.
2 The question of gender is relevant here and indeed throughout the filière. At the grower level most operations are 'joint' conjugal undertakings. This becomes less so with increasing scale. At the level of the integrators and processing different segments of the processing plant are highly gendered, though both men and women (and minorities in particular) figure centrally.
3 See for example the activities of United Poultry Concerns and the work of Davis (1995).
4 Animal and welfare rights questions relating to lesions, high mortalities, cannibalism, 'learned helplessness' and respiratory disease are profound. See www.iatp.org/hogreport
5 Murphy shifted to large scale confinement in the early 1970s and by 1995 Murphy Family Farms was the largest hog producer in the United States with sales of over $300 million.
6 This is documented in excruciating detail for an industry that has a long record of such egregious behaviour by the Sierra Club in Corporate Hogs at the Public trough (www.sierraclub.org/factoryfarms/).
7 Farrow-to-finish costs of production are now lowest in these so-called 'Prairie Gateway' states (see www.ers.usda.gov/data/CostsandReturns/car/hogs2.htm); Iowa appears as a relatively high cost state, above the national average.
8 Annual hog production reached a record high of 105 million in 1998 and prices paid to farmers fell dramatically. With excess capacity packers absorbed the

oversupply and posted record profits as independent growers saw prices fall to the lowest level since the Depression.

9 The deployment of the term here is simply a way of referring to new geographical divisions of labour, the creative geographical destruction by which one area loses a commodity and another gains. This is in contrast to the usage of the same term by Crang (1996).

10 Firms in both hogs and chicken while successful at commodity differentiation have been relatively unsuccessful in developing brand names and niche markets. The southern Californian 'Rocky Range Chicken' for example developed by Foster farms has not been successful, perhaps because as one reporter found the free range chicken gets two square feet of space to eat, drink and sit down.

11 The total elapsed time to accomplish a complete recycle of the entire breeder hen flock is about 520 days (the hog is at least twice as long). Broilers also grow quicker than hogs (11 weeks versus 9 months) and this grants more flexibility and a great opportunity for genetic improvements.

References

AER (1996) *Farmers' Use of Agricultural Contracts*. Washington, DC: USDA, Report 747.

Barboza, D. (2002) Chicken well simmered in a political stew. *New York Times*, 1 January, C1.

Berger, J. (1980) *About Looking*. New York: Pantheon.

Boyd, W. (1998) The real subsumption of nature, paper delivered to the Systems and Trajectories of Innovation in Agriculture, University of California, Berkeley, April 23–25.

Boyd, W. (2001) Making meat. *Technology and Culture*, 42, pp. 56–89.

Boyd, W. and Watts, M. (1997) Agro-industrial just-in-time. In Goodman, D. and Watts, M. (eds), *Globalizing Food*. London: Routledge.

Bugos, (1992) Intellectual property protection in the US chicken breeding industry. *Business History Review*, 66, pp. 127–68.

Clare, J. (1967) Love Hearken the skylarks. In Robinson, E. and Summerfield, G. (eds), *Selected Poems and Prose of John Clare*. Oxford: Oxford University Press.

Clare, J. (1982) The Mores. In Barrell, J. and Bull, J. (eds), *The Penguin Book of English Pastoral Verse*. London: Penguin.

Crang, P. (1996) Displacement, consumption and identity. *Environment and Planning A*, 28, pp. 47–67.

Davis, K. (1995) Thinking like a chicken. In Adams, C. and Donovan, J. (eds), *Animals and Women*. Durham, NC: Duke University Press.

Drabenstott, M. (2000) This little piggy went to market. *Economic Review of the Federal Reserve Bank of Kansas City*, Third Quarter, p. 97.

Fisher, C. (1999) Animal welfare likely to be on the WTO menu. *Bridges*, 3/6, pp. 13.

Friedmann, H. (1994) The political economy of food. *New Left Review*, 197, pp. 29–57.

Giedion, S. (1948) *Mechanization Takes Command*. New York: Oxford University Press.

Goodman, D., Sorj, B. and Wilkinson, J. (1987) *From Farming to Biotechnology*. Oxford: Blackwell.

Greenblatt, S. (1991) *Marvellous Possessions*. Chicago: University of Chicago Press.

Halverson, M. (2000) *The Price we Pay for Corporate Hogs*. www.iatp.org/hogreport/

Haraway, D. (1991) *Simians, Cyborgs and Women*. London: Routledge.

Harvey, D. (1998) The body as an accumulation strategy. *Society and Space*, 16, pp. 401–21.

Hobsbawm, E. (1994) *The Age of Extremes*. New York: Pantheon.

Horwitz, R. (1998) *Hog Ties*. New York, Macmillan.

Kautsky, K. (1899) *La Question Agraire*. Paris: Maspero.

Kleibenstein, J. and Lawrence, J. (1995) Contracting and vertical integration in the United States pork industry. *American Journal of Agricultural Economics*, 77, pp. 1213–18.

Knoeber, C. (1989) A real game of chicken. *Journal of Law, Economics and Organization*, 5, pp. 272–91.

Koonce, B. and Thomas, J. (1989) Differentiating a Commodity. *Planning Review*, 17, pp. 24–9.

Malamud, R. (1995) *Reading Zoos*. New York: New York University Press.

Martinez, S. (1999) *Vertical Coordination in the Pork and Broiler Industries*. Washington, DC: ERS/USDA, Report 777.

Martinez, S. and Reed, A. (1996) *From Consumers to Farmers*. Washington, DC: USDA, Report 720.

Martinez, S., Smith K. and Zering K. (1997) *Vertical Coordination and Consumer Welfare*. Washington, DC: ERS/USDA, Report 753.

Mayda, C. (2001) Community culture and the evolution of hog farming. In Flora, C. (ed.), *Interaction between Agroecosystems and Rural Communities*. New York: CRC Press, pp. 33–52.

NASS (2001) *Quarterly Poultry*. Washington, DC: USDA.

NPPC (1996) *Pork Industry Economic Review*. Washington, DC: National Pork Producer's Council.

Ollinger, M., MacDonald, J. and Madison, M. (2000) *Structural Change in US Chicken and Turkey Slaughter*. Washington, DC: ERS/USDA, Report 787.

Page, B. (1997) Remaking pork production, remaking rural Iowa. In Goodman, D. and Watts, M. (eds), *Globalizing Food*. London: Routledge, pp. 133–57.

Perry, J., Banker, D. and Green, R. (1999) *Broiler Farm's Organizations*. Washington, DC: ERD/USDA, Bulletin 748.

Polanyi, K. (1944) *The Great Transformation*. Beacon: Boston.

Pudup, Mary Beth (1983) *Packers and Reapers, Merchants and Manufacturers: Industrial Structuring and Location in an Era of Emergent Capitalism*. University of California, Berkeley, MA thesis.

Roenik, W. (1991) *Broiler Industry Vertical Integration*. Washington, DC: National Broiler Council.

Schlosser, E. (2001) *Fast Food Nation*. New York: Houghton Mifflin.

Thu, K. and Durrenberger, P. (eds) (1998) *Pigs, Profits and Rural Communities*. Albany: SUNY Press.

Tweeten, L. and Flora, C. (2001) *Vertical Coordination of Agriculture*. Washington, DC: Council for Agricultural Science and Technology.

USDA (2001) *Assessment of Cattle and Hog Industries 2000*. Washington, DC: USDA.

USITC (1992) *Industry Trade Summary: Poultry*. Washington, DC: Office of Industries.

Warren, D. (1958) A half century of advances in the genetics and breeding improvement of poultry. *Poultry Science*, 37/1, pp. 4–20.

Watt Poultry (2001) *Watt Poultry Journal* (annual). Atlanta, Georgia.

Watts, M. (1991) Life under contract. In Little, P. and Watts, M. (eds), *Living Under Contract*. Madison: University of Wisconsin Press, pp. 21–78; and Epilogue, pp. 251–63.

3   Spilling the beans on a tough nut

Liberalization and local supply system changes in Ghana's cocoa and shea chains

Niels Fold

Introduction

For most African countries, access to foreign exchange via loans has become limited due to the accumulation of foreign debt since the mid- to late-1970s. Loans from bi- and multilateral donors and international financial institutions are now linked to economic and political conditions for national economic policies, inscribed in so-called structural adjustment programmes (SAPs). In the agricultural sector this has involved a removal of subsidies on inputs and the liberalization of the domestic and external trade regimes. The latter previously had been heavily regulated because customs duties provided substantial income for governments of colonial and post-colonial African states. Liberalization and privatization of state owned or controlled companies involved in specific commodity sectors has been seen as a way of increasing competition through enabling private companies to purchase crops and sell inputs and services to agricultural producers (Gibbon, 2000). One of the main purposes of the SAPs has been to increase producer prices by removing costly state intermediaries between the world market and the individual agricultural producer and replacing them with competitive private traders.

This chapter examines how SAPs have affected the structure and dynamics of two agricultural supply chains in Ghana, namely cocoa and shea nuts. The central aims of the chapter are to map out similarities and differences between transformations in the two chains, and to explore the significance of convergence in the organizational structures of the local part of global commodity chains, as well as alterations to quality control systems. The restructuring of the two commodity chains is part and parcel of the gradual dismantling of the most important marketing board in Ghana – the former Cocoa Marketing Board (now known as Cocobod) – which previously also controlled the marketing and exports of shea nuts and coffee.

If the continuing liberalization process also removes the existing public quality control institution in the cocoa chain (a subsidiary of Cocobod), experience from other West African cocoa exporting countries suggests that no other institution – either private or public – will likely emerge to replace it. Instead, the most likely result will be a form of 'private global regulation'. Large scale cocoa grinding companies already have begun to implement their own quality control systems at the point of export, that is, sorting and grading at warehouses in the major harbours. However these systems do not comply with an emerging set of process standards related to food safety concerns in the EU, nor do they fulfil claims from NGOs, consumer organizations and media in the North requiring certainty for ethically acceptable working conditions as has been the case in other global commodity networks (Hughes, 2001). One way to deal with both concerns is through the incorporation of some kind of traceability in the organization of the local supply chain. Yet this is an extremely complex task in chains completely dominated by smallholder cultivation of what traditionally has been considered as a bulk product. Although substantial elements of traceability are currently incorporated in the existing quality control system in the cocoa chain, this system is likely to be dismantled as the liberalization process proceeds.

Both commodities are used in the food industry, particularly in that part of the confectionery industry which manufactures products with chocolate content. Cocoa is by far the most important and well-known ingredient in chocolate whereas vegetable fat from shea nuts (mixed with other vegetable fats) has served as a little known auxiliary ingredient (cocoa butter alternatives) in chocolate production in countries such as Ireland, the United Kingdom and Denmark for a number of years. In other nations such as France, Belgium and The Netherlands, chocolate with a content of vegetable fat other than cocoa butter traditionally has been considered as a low quality product. Differences both in national taste preferences and in food legislation are socially constructed and conditioned by a fascinating mix of cultural, economic and political factors (Fold, 2000).

The first section of the chapter outlines the technical and political inter-linkages between the two commodities that – to a certain extent – simultaneously supplement and compete with each other on the world market. I then examine the present organization of the large-scale trader segment, the structure of the purchasing systems and quality control practices within each of the two chains. Finally, I summarize structural similarities and differences between the two chains and reflect upon the consequences of introducing new process standards linked to traceability of future commodity exports.

Unless otherwise stated, details concerning the restructuring process originate from ongoing research on globalization and economic restructuring of perennial crop chains in Ghana. For the purposes of this chapter, my primary concern is with the structural composition and relationships between producers (pickers), local traders and exporters in Ghana. The chapter touches only briefly upon the logistics, financial flows, and links to the industrial customers and private consumers of finished goods in the North, and does not include issues related to the manufacture and exports of processed cocoa beans or shea nuts from Ghana (see also Fold, 2000, 2001). Although local processing of both commodities currently is increasing, the trade and export of raw materials to industrialized countries, primarily within the EU, predominates.

Chocolate with 'nuts'?

Cocoa has been the most important cash crop in Ghana since it replaced palm oil in the beginning of the twentieth century. It is primarily cultivated in the deciduous forest areas of the southern part of the country (Figure 3.1). Due to careful harvesting and handling practices among Ghanaian farmers and a well-functioning quality control system, Ghanaian cocoa beans have the best reputation of all so-called bulk beans used by the chocolate industry, primarily in Organization for Economic Cooperation and Development (OECD)-countries.[1] Because of this reputation, most cocoa from Ghana has been – and continues to be – sold on forward terms, that is, to buyers willing to sign contracts up to 12–18 months before the actual harvest of the beans. Following a sharp decline

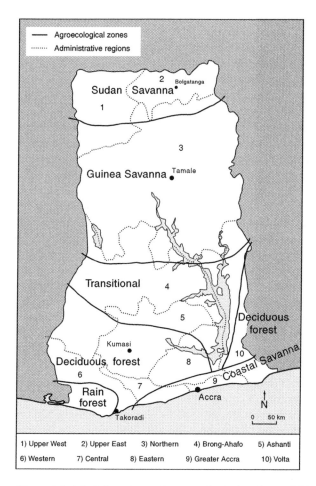

Figure 3.1 Administrative and agroecological regions of Ghana.

Source: Author.

in the 1980s due to a mixture of drought, mismanagement and political turmoil, cocoa production in Ghana increased substantially during the 1990s. The country is now the world's third largest exporter with a volume of about 400,000 tonnes, roughly the same size as Indonesia (the second largest producer) but far behind neighbouring Cote d'Ivoire, the largest producer of cocoa with an export of about 1,200,000 tonnes (ITC, 2000). On a global scale, West Africa now has regained and strengthened its position as the dominant producer in the region following the serious decline of the sector in countries with plantation-based production. In Brazil problems stemmed from a widespread plant disease and in Malaysia land increasingly has been given over to other perennial crops, notably oil palms. In West Africa, cocoa production is almost completely carried out by smallholders although the extent to which some farmers have accumulated user rights to a large number of relatively small cocoa-fields is not publicly known.

Shea nuts originate from trees growing wild in the southernmost part of the Sahel and the adjacent Sudan and Guinea savannas in West and Central Africa (von Maydell, 1986). In Ghana, however, trees are scarce in the southern part of the Guinea Savannah. Current tree population levels (10–50 trees/hectare) derive from human selection, including protection from cutting (charcoal production) and fire during the cyclical clearing of fallow (CABI, 2000). Trees are not commercially planted and cultivated in plantations because of their slow growth (they take up to twenty years to mature) and because nuts are a highly variable crop both in terms of yield and quality. Consisting of a kernel covered by a shell, shea nuts are found within plum-size green fruits. The nut contains a vegetable fat (shea butter) which is used locally as a frying medium or added to various sorts of porridge.[2] The nut also can be transformed into soap by adding wood or plant ashes to the boiling mixture of fat and water (UNIFEM, 1997). The annual volume of exports from Ghana is estimated to have peaked at about 20,000 tonnes in the mid-1990s (Cocobod, 2000) although actual annual exports fluctuate quite dramatically due to climatic variations and the natural cycle of the trees. Figures for total exports from the West African region are very uncertain but it is estimated by Obi (2000) that exports only constitute a minor part (10–15 per cent) of local consumption (about 550,000 tonnes). It is notable, however, that an unknown amount of nuts (estimated by Obi (2000) at about two times the volume consumed locally) remain uncollected when they drop to the ground.

In OECD-countries, shea butter is used together with other tropical (vegetable) oils in the production of cocoa butter alternatives (CBAs) which are used as a substitute for cocoa butter. The demand for shea nuts on the world market is determined by the demand of the chocolate and food industries for different types of cocoa butter alternatives. The chocolate industry in particular uses specific types of cocoa butter equivalents (CBEs) which possess similar physical and chemical properties to cocoa butter. On a global scale, only a handful of companies have mastered the technologically very advanced production of CBEs, namely Fuji Oil (Japan), Unilever (the United Kingdom/Netherlands),

Aarhus Olie (Denmark) and Karlshamm (Sweden). The use of CBEs in chocolate production has a number of technical benefits which improve the final product (in terms of bloom, gloss, snap, melting point, and shelf life). In addition, the production process is easier to control and the product specifications are more stable from batch to batch (Brun, 1998; Shukla, 1995). Cocoa butter also is generally far more expensive than CBEs, although the price relation depends upon the quality and current price of the cocoa beans and the specific CBE in question. In addition to the technical benefits CBEs provide, the replacement of cocoa butter with CBEs also allows chocolate manufacturers to cut production costs.

During the mid-1990s, the use of CBEs in chocolate production was subject to a heated public debate in the EU, the so-called 'European Chocolate War'. As mentioned above, different national traditions prevail in the way chocolate is produced and consumed in the EU. The use of CBEs in chocolate production is permitted in some countries but not in others. In countries where the use of CBEs in the manufacturing process is inhibited, CBE-containing products are not allowed to be marketed as chocolate. This variation in food legislation has been a problem for efforts to harmonize rules concerning food production and marketing in the EU since the entry of Ireland, the United Kingdom and Denmark in 1973. Agreement on common rules for production and labelling of chocolate products in the EU was only achieved after almost twenty-five years of discussion. New rules took effect in August 2003 which permit up to 5 per cent of vegetable fat in chocolate products provided that it is clearly stated on the wrapping.

The debate revealed a pronounced disagreement among northern development NGOs. NGOs working in the Sahel area stressed the socio-economic importance of shea nuts for female collectors (given that the export market is an important alternative to local markets) as well as the fact that income from sales of shea is directed towards the needs of the household. Other NGOs linked to target groups in cocoa growing regions argued that permission to use cocoa butter alternatives would substantially reduce the market for cocoa, resulting in decreasing prices and incomes among (male) cocoa farmers. Governments of the cocoa producing countries in West Africa were opposed to the new regulation for the same reasons. Most of the countries, however, also export shea nuts although the economic importance is relatively insignificant in comparison with that of cocoa. During the first half of the 1990s, the value of average annual exports of cocoa and processed cocoa products from Ghana was US$321 million constituting about 35 per cent of total exports (ITC, 2001) while annual exports of shea nuts were less than US$5 million (Cocobod, 2000). Economic and political pressure groups linked to the national shea-chain are substantially weaker than groups linked to the cocoa-chain and the official position of the Ghanaian government was (apparently) never seriously challenged in the public debate. The different positions have potentially explosive ethnic and religious undertones, as cocoa cultivation is located in primarily Christian parts of the south whereas shea nuts are picked and processed in the Muslim north.

Restructuring of the large scale trader *cum* exporter segment

Shea

Prior to the liberalization of domestic and external trade in shea nuts during the early 1990s, activities in the chain were regulated by the Cocoa Marketing Board (CMB) who purchased the nuts through its subsidiary, the Produce Buying Company (PBC). The PBC traded with local licensed traders and/or farmers' societies, organized stocks and sold consignments to overseas industrial customers, partly via international traders (Chalfin, 1996).

In the first few years after liberalization, the PBC gradually ceased to operate in the shea nut trade and the company has not been involved in the chain since the mid-1990s.[3] A number of small local traders expanded business as independent companies, many of them already holding a license to buy nuts for PBC. New and somewhat larger trading companies also began to operate, largely staffed with purchasing clerks and managers previously employed by the PBC. The companies included some of the former state-owned agricultural input suppliers, who were squeezed by emergent competitors as their former monopolistic position changed under liberalization. However, none of the new companies have survived. A main reason for the collapse of 'would-be' large-scale trading companies has been their inability to organize the purchase, transport and sale of the nuts at reasonable efficient speed and with sufficient security of capital lay out, for example by entering into risky investments or pre-financing arrangements.

Some CBE manufacturers also attempted to establish subsidiaries at the beginning of the liberalization process but none are now directly involved in buying operations in the upper part of the supply chain in Ghana. At least one of the companies, however, has established a small exporting company in Accra in order to monitor and organize storage facilities and shipping of shea nuts from the Tema harbour.

Only two companies are now active in the large-scale shea-trade in Northern Ghana, namely an international trading company and a local company.[4] The international trading company is a subsidiary of the Nanrai group of companies, an international trading company of Indian origin. Local subsidiaries in various African countries are linked to Olam International, a specialized commodity trading company located in Singapore. The most important activities for Olam in Ghana are imports of rice and sugar and exports of shea nuts, coffee, cashew nuts and cocoa. The company is able to buy throughout the whole crop season due to its relatively easy access to credit, either from external sources or supplied from other divisions in the Ghana subsidiary: capital can be transferred to cover transactions in a specific commodity from other trading activities. Olam entered the shea trade quite rapidly by hiring selected PBC-employees, administrative as well as purchasing personnel.

The local company, Kassardjian, has operated in Ghana since the significant shea-trade started. The company is owned by an Armenian family (with roots in

Lebanon) who also has considerable interests in Accra (construction, hotels, etc.). While constructing roads in Northern Ghana for the British colonial government, the founder became aware of the ample presence of shea nuts. In the 1950s, samples of shea butter were sent to the United Kingdom for testing and exports increased rapidly in the following years, both of nuts and butter. After independence when the trade in nuts was taken over by the CMB, the company operated as a licensed buyer and sold the nuts to PBC – and repurchased them for exports to customers in Europe. The company in Ghana also trades in coffee and has established sister-companies in Benin and Togo which export the same commodities. All marketing and sales operations are managed by the company's office in London.

Cocoa

Until 1992 the purchase, storing, quality control and shipping of cocoa beans was carried out by various subsidiaries of the CMB. The purchasing arm of the marketing board, the PBC, was the only company allowed to buy and store beans from farmers. The Quality Control Division (QCD) was exclusively involved in grading and sealing of the beans and the Cocoa Marketing Company (CMC) organized all exports and shipping of Ghana cocoa. In addition, supplies of various inputs for cocoa cultivation, extension services and cocoa-related research were solely handled by other subsidiaries. Transport (secondary evacuation) of cocoa beans from the upcountry depots to the ports (Tema, Takoradi and the 'in-land' port Kaase near Kumasi) was partially outsourced to private haulage companies, partially carried out by PBC itself using company owned vehicles. The transport from cocoa buying posts in the villages to the PBC-owned depots (primary evacuation) was more or less fully taken care of by PBC although local hiring of pickups or small lorries occurred in situations of transport deficiency.

Because liberalization and privatization were initiated at the start of the structural adjustment programme in the agricultural sector, the structure and dynamics of the cocoa chain in Ghana have been changed but not fundamentally transformed. Supplies of inputs (fertilizer, pesticides, etc.) have been privatized at the same time as subsidies were removed and extension services to cocoa farmers have been merged with other extension services in the Ministry of Agriculture. However, the QCD is still the only body with exclusionary rights to grade and seal the cocoa beans and the CMC is still the sole exporter of cocoa beans – even though changes have been introduced since 2002. Moreover, the pricing system is maintained so that a pan-seasonal, pan-territorial producer price is made public by Cocobod (the restructured successor of CMB) before the main crop season officially commences on 1 October. Prior to the public announcement, all margins for the services involved (quality control, transportation from depots to ports, storage and marketing) are determined and negotiated between representatives from the Finance Ministry, farmers and the various business groups.

More substantial changes, however, have occurred in the purchasing segment of the chain where private companies have been allowed to operate since adjustment programs began. Gradually these companies have taken over a larger share of total purchases so that the PBC now constitutes only about 40 per cent of total volume purchased. Most of the companies are locally owned but three new actors recently have entered the cocoa business: the international trading company Olam (mentioned earlier), a joint venture between foreign and local interests, and a fully foreign owned company. All three companies have set up the necessary infrastructure (buying posts, depots, trucks and offices) although at the time of writing the fully foreign owned company still had not obtained a purchasing licence from Cocobod. In addition to the main purchasing companies (ten in the 2000/2001 main crop season, each having purchased more than 1 per cent of total intake since liberalization started), a number of smaller companies are officially listed as licence holders. In practice these smaller companies have not purchased significant volumes of beans in the period. Although some were quite active in the past, they have now diverted their business into other areas or remain dormant, waiting for further capital injections.

It is a notable feature of the local private companies that some of the largest are owned, fully or partially, by business groups with a long standing interest in the lucrative transportation between the depots and the ports as well as in pre-shipment storage of beans. Accustomed to the PBC's operations upcountry, and apprehensive of the risk of being cornered or pressured by new cocoa procurement companies, these major haulage companies have formed their own purchasing companies, particularly in the late 1990s. One of the haulage companies has created four independent companies ('GR4' in Figure 3.2) with different partners, which are all active and quite successful.

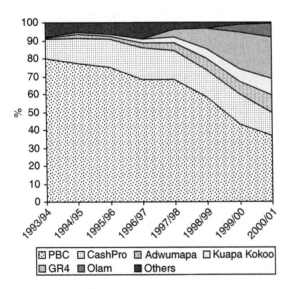

Figure 3.2 Market shares (%) of LBCs, 1993/94–2000/01.

Source: Cocobod (unpublished data).

Flexibility and rigidity in local purchasing systems

Shea

The shea purchasing system consists of female pickers, local agricultural wholesale traders (of varying size) and independent agents supplying the relatively resource rich trading companies. The latter employ their own purchasing agents and as mentioned above, also organize exports of the nuts. A myriad of different relationships link female pickers with the local trading segment. Some traders, both male and female, seek out societies, that is, groups of female pickers in or around a village, buy the available dried nuts and organize transportation of nuts to their home town. Others base themselves in towns and buy from arriving pickers/ farmers, in particular on market days; some also may buy nuts from the first group. Depending on their financial position and liquidity requirements they will sell nuts to large-scale trading companies or store them in expectation of higher prices late in the season. The offer price of the large-scale traders is based on the selling price to overseas industrial customers, taking into consideration what competing companies are offering and – to a lesser extent – the local demand of nuts for household processing and consumption. Usually prices are adjusted upwards during the season because of local supply and demand conditions and sharper competition between exporting companies, who aim to fulfil their contracts with overseas buyers.

During the 1990s, the two exporting companies developed a structurally similar pattern of organizing their purchases of nuts. A crucial feature of the arrangement is the high degree of risk involved in pre-financing the individuals who conduct the actual buying operation. Independent traders are usually not pre-financed although some of the more trusted ones sometimes are given cash for specific trading operations – or when nuts are scarce, prices high and competing trading companies offer pre-financing. Independent traders have long-established networks that link them to a number of communities or villages from which they purchase the nuts. Sometimes it is possible to obtain nuts on credit terms but villagers also may require cash at delivery, particularly if prices are high or more buyers are operating in the area. Usually the traders are paid a commission 'per bag' (i.e. based on weight).

In contrast, cash is provided to company employees (purchasing clerks) on the basis of procurement records for previous years but only in amounts that are estimated to cover purchases for about a week. If a clerk has identified larger quantities of nuts for sale, he must request more cash. Company employees travel into more remote areas than independent traders in order to source nuts from more isolated communities. Nuts from these places often are used for subsistence or stored until means of transportation becomes available. If sufficient volume of nuts are located in a remote village, the purchasing clerk has the opportunity to earn income in addition to his salary by buying at a lower price and organizing the transport to company depots.

A number of agents/brokers operate at markets in larger towns, particularly in Tamale. Most agents are connected to one of the large-scale trading companies but also act as brokers for competitors if the opportunity arises. On the supply side, each agent is linked to a handful of large traders in the Tamale market although they also buy from smalltraders based in smaller towns, especially on market days when small traders come to Tamale. Immediately (i.e. normally within the same day) after sufficient volumes have been purchased on order from the customers, nuts are transferred to the large-scale trading company as agents do not take possession of the nuts, nor do they have warehouse facilities at their disposal.

Cocoa

The post-liberalization practice of the private cocoa buying companies (called licensed buyer companies – LBCs) do not differ much from company to company. This uniformity can to a large extent be ascribed to the fact that most LBC managers are former employees of the Produce Buying Company, as well as to the fact that the fixed price system has been maintained. However, operational terms, management structure and spatial organization may differ. Some companies operate with a regional coordination of activities at a level similar to administrative regions while others operate with 'units' or 'sectors', that is, a smaller group of districts not making up a whole region. Smaller companies tend to have no intermediate coordination between districts and the head office. Apart from this, the basic organization follows the traditional outline from the PBC-monopsony although with some notable differences.

Before liberalization, the starting point was constituted by a local buying post (shed), staffed with a purchasing clerk who bought cocoa beans from farmers. The PBC predominantly hired male farmers on an annual basis as purchasing clerks but females and non-farming males also were employed. The farmers themselves provided for transportation from farm gate to the nearest PBC-shed, often carrying beans in sacks or bowls. At the PBC-shed beans were weighed and (re-)bagged in 64.5 kg sacks.

Since liberalization, purchasing agents have been hired predominantly on a commission basis rather than as employees with fixed annual salaries: agents are paid a sum of money for each bag of cocoa they deliver to the company. In most LBCs, however, proven individual ability to provide a substantial volume of purchased beans per season will qualify the agent to an annual salary. Hence, the status of a purchasing agent will be similar to that of a purchasing clerk in the former PBC-system. Agents (or clerks) may organize transport from large farmers or remote societies at their own cost if they think a profit can be earned. Likewise, they can hire people to sort, dry and bag the cocoa they have purchased if they choose to devote their time to other tasks.

Usually farmers are paid in special Akuafo cheques that – in principle – can be cashed or credited to the farmer's personal account in the local branches of national banks. The system was introduced in the 1980s to reduce incidences of cheating by purchasing clerks and to expand the network of rural banks (Gyimah-Boadi, 1989). However, some farmers, particularly those living in remote areas, prefer payment in cash rather than wasting time and money by going to town and waiting in banks. Not only do rural banks often have liquidity problems but also LBCs frequently may not transfer sufficient money to cover all claims from farmers. As a result, some of the LBCs have introduced payment in cash as a competitive parameter, although many farmers still prefer payment in Akuafo cheques because regular registration as a cocoa-selling farmer is a means to enter special state-financed agricultural credit schemes.

As many more companies become active in most of the cocoa producing areas, prompt payment via cash or cheque is increasingly important for a company's ability to attract farmers and thereby stay in business. Farmers increasingly 'shop around' now between different company agents located in the same village, selling the beans to those who are able to pay.

The introduction of cash requires extremely close monitoring of agents and clerks in order to increase the security of funds flowing between the company and its employees. In comparison with normal rural income levels, the amount of money handled by a clerk is significant. Community or family pressures on the local clerk for short term credit may also result in company losses as the loans seldom are repaid. Thus, although the new competitive system reduces company costs to salaries and introduces incentives for the employees to increase purchases, it also increases the company's dependence on the competence and integrity of the agent or clerk. As a consequence, the selection of new purchasing agents is carried out with great care. Apparently young men are preferred as agents by the LBCs. The chief and elders are asked to recommend suitable candidates and company representatives then carry out interviews with those short-listed. Candidates must be literate and able to socially engage with fellow villagers so that cocoa is brought to them rather than to competing purchasing companies. Moreover, agents must be able to provide a collateral in the form of a house or other kind of property, usually secured via assets owned by their family.

In some cases, special characteristics are highly valued, for instance the ability to target specific groups by having a certain ethnic, political or religious affiliation. This issue is particularly relevant in the western region which is Ghana's biggest and most important area for cocoa production, constituting about 50 per cent of total production since the mid-1990s (Cocobod, n.d.). Although cocoa production has existed in this region for many years, current production has vastly increased in scale. Many farmers are migrants from other parts of Ghana resulting in a broad variety in ethnic and religious background among and within local communities.

Quality control systems and product standards

Shea

A distinctive mark of the shea sector in Ghana is the almost negligible role of quality requirements and the absence of (measurable) standards in some of the trading transactions in the chain. The large-scale trading companies buy nuts with little quality control beyond a cursory visible inspection of the extent of mouldiness. All colours and shapes of nut are accepted and bought at the same price although the industrial customers pay a premium for nuts with a low moisture content and free fatty acids – or reduce the price accordingly for a high content. Quality is usually checked by an independent inspection company in Tema before shipping and again at the reception at the factory gate in the north. This lack of attention to nut standards (as a quasi measure of quality) is strikingly different from the ways in which local traders deal with pickers or other traders (Chalfin, 2000). Most of those involved in the local food trade know that black nuts have been exposed to rain or open fire in the drying period and that the butter made from them is of a lower quality due to higher content of free fatty acids. In seasons where nuts are abundant, quality parameters such as these are used to determine price levels in local transactions.

Cocoa

The quality control system in the cocoa chain remains more or less intact compared to the pre-liberalization system but sharply contrasts with the practices in the shea nut trade. Under the old system, the district division of the QCD was 'invited' to inspect beans when the purchasing clerk considered the volume of purchased beans to be sufficient. Depending upon existing workload levels, the QCD would send several graders to check the quality, grade the beans and seal each sack. Present practice is still in line with this very protracted procedure: samples are taken from each sack in one lot (i.e. a collection of thirty sacks) and mixed. From this collection of beans, a sample used for grading is picked out. After visible inspection for oddly formed beans, mouldiness and other defects, beans are tested for moisture content and a 'bean count' (i.e. number of beans per 100 grams) determines whether the beans are of an acceptable average size. Lastly, a laborious 'cut test' is conducted. This test consists of 100 beans taken from the sample, sliced into halves and checked for mouldy and slaty (under-fermented) beans. The extent of the former affects the flavour of the cocoa liquor whereas the latter influences the quality of the cocoa butter. Cocoa beans are either accepted as exportable grade 1 or grade 2 beans or are rejected as substandard beans and are taken away from the shed for re-conditioning and re-sorting.[5] The same producer price is paid for both grades in order to avoid cheating farmers – although normally almost all beans are of grade 1. After passing quality control and grading, each bag of cocoa is sealed in order to identify the QCD grader, the date and the specific shed from where the beans originate.

The only new element in the quality control system is logistical in nature. Due to pressures from the LBCs who had no substantial local storing facilities in the early phase of their operations, the quality control system has been changed so that grading and sealing no longer takes place at the shed level but rather only in the depots. Furthermore, all depots now must be located near the cocoa district capital, a rule that reduces QCD's operational costs and time consuming transport between sheds.[6] These cocoa districts differ from political-administrative districts; they are functionally separated entities set up in cocoa producing regions. The clear loser in the new system is the PBC which prior to liberalization owned several thousand sheds, including many relatively large ones. Many of these, however, are located in smaller towns or villages and cannot be fully utilized under the present system.

When grading and sealing have been completed and an adequate number of bags have accumulated in the depots, the beans are transported to the ports and taken over by the CMC if the consignment passes a similar quality control procedure conducted yet another time by the QCD. A third and identical quality control procedure is implemented immediately prior to shipping.

Liberalization, organizational changes – and new process standards

Although many new trading companies operated during the initial liberalization period in Ghana, the total shea export trade now is controlled by only two companies. Close to 85 per cent of total cocoa exports are domestically purchased and controlled by six LBC-groupings. The foreign owned company in the shea chain is now an increasingly important actor in the cocoa-chain and other foreign companies have become involved in the cocoa trade. It remains to be seen whether the operations of these foreign-owned companies will differ from the present generalized practice. The importance of the former state-controlled company has been substantially reduced in the cocoa chain and has been removed completely from the shea chain. Local companies who began operating in the cocoa chain during the mid- and late-1990s still maintain a relatively strong position in the purchasing segment, particularly those that also operate in the secondary evacuation segment.

Most likely, new linkages between the local LBCs and exporters or captive buyers will emerge, structurally similar to those in the shea chain (see Figure 3.3). The long-term aim of the structural adjustment programme is to liberalize exports of cocoa exports completely but the LBCs are presently only allowed to export 30 per cent of their cocoa intake directly to customers. So far, however, all the companies have voluntarily sold all their cocoa to the CMC which then covers forward contracts or sells the cocoa on the spot-market. The scope of local company operations and the volume of capital are still too small to use the futures market. Without hedging possibilities, trading on the world market is too risky and the local LBCs need to find captive buyers amongst cocoa grinders. These new relationships pave the way for possibilities to enter into contractual relations including some kind of pre-financing of purchasing activities, in turn reducing the LBC's dependence on the CMC for this facility. Similar linkages can be established to commodity exporters and/or chocolate companies that may wish to import cocoa beans directly from the country of origin (Fold, 2001). The fact that foreign trading companies are establishing purchasing operations within the chain reflects current constraints on local companies. Foreign companies have the necessary organizational and financial capital to 'play' the futures market for cocoa and to deal directly with grinders (and chocolate companies) in the same way that the two large-scale trading companies in the shea chain deal directly with CBE manufacturers.[7]

In both systems, crucial parameters for the total profit of the large scale trading companies are the volume of working capital and the speed of capital turnaround time – that is, the time used for the circulation from money to commodity capital and back again. Flexible logistical and payment systems for purchase and primary

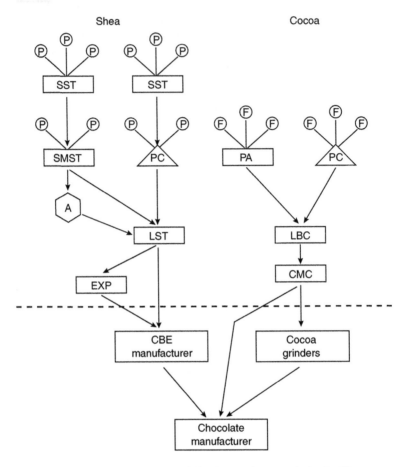

Figure 3.3 Organizational structure of the shea and cocoa chains in Ghana.

Source: Author.

Abbreviations: A: agent, CMC: Cocoa Marketing Company, EXP: exporter, F: farmer, LBC: licensed buyer company, LST: large-scale trader, P: picker, PA: purchasing agent, PC: purchasing clerk, SMST: small and medium-scale trader, SST: small-scale trader, CBE: cocoa butter equivalent.

evacuation are therefore required. Although the purchasing systems of the two chains seem to differ there are reasons to believe that a continued liberalization of the cocoa chain would result in an expansion of the initial trading segments. A dismantling of the fixed price system for cocoa would stimulate an extension of the hierarchical system among small and medium traders and would create space for agent-like activities to coordinate and assist the operations of the LBCs. Thus there are reasons to believe that the purchasing system in the cocoa chain would structurally converge towards the shea chain (Figure 3.3).

Each chain tends to be based on a core group of salaried purchasing clerks covering an important part of the sourcing strategy of the large scale trading company as they operate in areas where commodities are plentiful but perhaps not easily accessible. In the case of shea, the purchasing clerk controls limited funds for buying nuts whereas the clerk in the cocoa chain normally uses the cheque system or pays cash with company money. Both chains operate with a group of more loosely linked traders or agents although this is more evident in the shea chain than in the cocoa chain. In the former, the relationship is market-based although some of the small- and medium-sized traders indicate the existence of relatively stable links between them and one specific large-scale trader. In the latter, purchasing agents are formally linked to one company by a modest collateral and relations of 'trust' because the purchasing agent handles relatively large sums of money, either in the form of the right to issue cheques or pay in cash.

Due to the maintenance of the state-monitored quality control system in the cocoa trade, the role of quality conceptions and standards are very dissimilar in the two chains. It is a peculiar difference that quality considerations are important in the local shea nut trade but not significant in the export trade. In contrast, standards adopted in accordance with quality conceptions among Western European chocolate manufacturers close to fifty years ago are still maintained despite bearing absolutely no relation to local preferences and demand. Clearly, this situation relates to different use values of the commodity as a local food ingredient in contrast to a sheer cash crop outside local diet patterns. In addition, this difference between chains is embedded in gender relations as women control and organize the purchase and processing of shea nuts for butter whereas it is usually men who control the cocoa harvest.

Perhaps surprisingly, debates surrounding quality in the 'Chocolate War' among consumers, retailers and cocoa processors (including grinders and chocolate manufacturers) in the north did not affect the organizational structure and dynamics in the upstream parts of the chains in Africa. However, conditions in the producing and exporting segments rather than 'taste-preferences' in the consumer segment have been moved to the epicentre of the most recent public debate involving the chocolate industry. Children were discovered living and working under slave-like conditions on cocoa farms in Cote d'Ivoire, and the British and American media featured stories about children who were lured or 'sold' by parents or guardians in the Sahelian countries to unscrupulous cocoa farmers further south.[8] Here children were provided with miserable food and accommodation and forced to labour in the fields without any payment. A series of television and newspaper stories resulted in a public outcry of moral indignation: in the United States several members of Congress demanded immediate action by the chocolate industry.[9] After initially rejecting the claims, the industry reluctantly admitted the existence of possible problems upstream in the supply chain. Large-scale companies entered a voluntary agreement to investigate the scope of the problem in West Africa, including adopting a scheme which would guarantee 'slave-free' cocoa beans. Large-scale companies in the European chocolate industry, organized in the organization CAOBISCO, also are involved in the initiative. Ultimately, the scheme will be based in some way upon the principle of traceability.

In fact, this initiative combines ethical virtues with commercial necessity. From 2005, traceability will become a legal requirement for all ingredients used in food and feedstuffs' production within the EU.[10] The aim of the legislation is to incorporate various sanitary requirements linked to different stages in the production and distribution process in order to secure food safety at the terminal point of consumption. In the process, an emphasis upon quality control shifts from product to process standards which in turn requires an organizational set up that is able to accommodate traceability requirements. This will cut against the completely different situation for African smallholder produced bulk commodity exports that are still traded according to product standards. Not only will traceability be required within a very short time horizon, but also the global chocolate industry faces almost insurmountable problems in determining 'origins' within a raw material supply which is highly dependent on innumerable and anonymous West African small-scale cocoa farmers. Even more complex is the task of introducing traceability in the shea nut chain, in which semi-wild tree products are picked according to traditional local rules about access to natural resources.

The incorporation of new process standards based on traceability is likely to be even more difficult to achieve in Ghana after full-scale liberalization according to the structural adjustment ideals of the World Bank. Traceability in the present Ghanaian system reaches upstream as far as the local buying post: the high quality of cocoa beans is due to the implementation of control measures almost at farm gate level. Yet the experience from other cocoa exporting countries in the region indicates that liberalization has resulted in the dismantling of existing quality control systems without the establishment of alternative systems (Fold, 2001). Local traders seem to prefer shorter capital turnround time at a lower unit price rather than to maintain a quality-linked premium that would involve laborious and time consuming control procedures. Until now, it also has been possible to sell lower quality beans to Northern cocoa grinding companies.

Conclusion

Structural adjustment in Ghana's perennial crop sector has tended to stall competition in the large-scale trader segment whilst stimulating a trend towards uniformity of local purchasing systems within and across the two commodity chains. Despite variations in the extent of liberalization of the chains, existing quality control systems and product standards have been relatively stable. However continued liberalization of the cocoa chain in Ghana will most likely lead to the disappearance of the cocoa quality control system as has been the case elsewhere in West Africa among cocoa producing countries. Ironically, this situation has been created at exactly the same time that (some) northern consumers have begun to make increasingly vocal calls for the monitoring and traceability of commodities. New process standards based on traceability will be difficult to introduce in the institutional vacuum left over from the former system – not to mention the problems of implementing the principle in a non-commercialized tree crop chain like shea. The long-term effects of failure to implement traceability in West Africa would be dramatic for small-scale farmers and national economies in West Africa as global demand for cocoa would gradually be directed towards large-scale plantations in Brazil and Malaysia with the ability to ensure traceability requirements.

The chapter has highlighted a lack of cohesion between neo-liberal adjustment policies and new forms of process standards on food products consisting of small-holder crops from developing countries. The implications of this disjuncture are significant. In the absence of monitoring and regulation of quality control systems by local public institutions, large-scale grinders and chocolate manufacturers in the north may seek to vertically integrate upstream in local supply chains by setting up elaborate and costly intra-company quality assurance schemes based on their ability to monitor and regulate cultivation, purchasing and logistics (see Morris and Young, Chapter 4, this volume). This would mark the final stage of the reversal to private regulation from state controlled cocoa exports of West Africa that was founded in the late colonial period and developed by independent African governments before the implementation of SAPs. Whether the power dynamics of this new system of regulation will mirror older colonial ties is an open question. Undoubtedly, relationships will be shaped by the willingness and capability of northern consumers to incorporate demands for decent living and labour conditions for small-scale African cocoa producers in traceability requirements, even if that requires paying a higher price for satisfying their sweet tongues.

Acknowledgements

I would like to thank Suzy Reimer and Alex Hughes for the inspired editing of initial drafts. I am also grateful for valuable comments from Jim Pletcher and Peter Gibbon.

Notes

1 'Fine and flavoured cocoa' from the Caribbean and Ecuador is much more expensive but it forms a quantitatively minor part (about 5 per cent) and a distinctive segment of the world market for cocoa.
2 In fact it is the shea kernel that contains fat but most actors in the chain use the word 'nut' synonymously with 'kernel'. For reasons of simplicity I use the same terminology in the remaining part of the chapter.
3 Rumours in late 2001 said that the company considered to start operations again in the 2002 season and that representatives tried to mobilize the old supply-network.
4 In late 2000, a former manager from the local company established his own company and started operations at the beginning of the 2001-season, exploiting his personal experience and well-established connections with local traders and industrial buyers.
5 Substandard beans that failed the re-sorting are used for animal fodder or sold at a discount to a specialized foreign-controlled company (Resigha) that exports them for processing at their factory in Amsterdam.
6 This is the reason for a distinctive new urban structure in the cocoa district centres, especially in the major producer regions where most of the companies are active. New depots of similar design but of varying size are being built outside the town centres, close to the main approach roads from the farming areas.

7 The processing of CBE is highly specialized and the ingredient is considered of less vital importance for the 'taste' of the chocolate than cocoa liquid and cocoa butter. Hence, all chocolate manufacturers are externally supplied whereas some of them for various reasons still process part or all of their cocoa-based raw materials (Fold, 2001).

8 See, for example, www.truevisiontv.com

9 See www.krwashington.com

10 Regulation (EC) No 178/2002 of the European Parliament and of the Council (of 28 January 2002) published in Official Journal of the European Communities 1.2.2002 (L 31/1).

References

Brun, G. H. (1998) Le beurre de karité dans le chocolat: l'etat de la question. In Sawadogo, S. *et al.* (eds), *Etat des lieux de la filière karité, gestion de la qualité, connaissances et savoir, marchés, cadre de concertation des intervenants de la filière*, Ouagadougou: Projet Filiere Karite, pp. 109–21.

CABI (2000) Vitellaria paradoxa [original text by P. Lovett]. In *Forestry Compendium Global Module*. Wallingford (UK): CAB International.

Chalfin (1996) Market Reforms and the State: the Case of Shea in Ghana. *The Journal of Modern African Studies*, 34(3), pp. 421–40.

Chalfin (2000) Risky Business: Economic Uncertainty, Market Reforms and Female Livelihoods in Northeast Ghana. *Development and Change*, 31, pp. 987–1008.

Cocobod (2000) *Ghana Cocoa Board Handbook*. Accra Ghana Cocoa Board.

Cocobod (n.d.) Unpublished data.

Fold, N. (2000) A Matter of Good Taste? Quality and the Construction of Standards for Chocolate Products in the EU. *Cahiers d'économie et sociologie rurales*, 55–6, pp. 92–110.

Fold, N. (2001) Restructuring of the European Chocolate Industry and its Impact on Cocoa Production in the West. *Journal of Economic Geography*, 1, pp. 405–20.

Gibbon, P. (2000) The Impact of Structural Adjustment. In Friis-Hansen, E. (ed.), *Agricultural Policy in Africa after Adjustment*. Copenhagen: Centre for Development Research.

Gyimah-Boadi, E. (1989) Policies and Politics of Export Agriculture. In Hansen, E. and Ninsin, K. A. (eds), *The State, Development and Politics in Ghana*. London: Codesria, pp. 222–41.

Hughes, A. (2001) Global Commodity Networks, Ethical Trade and Governmentality: Organizing Business Responsibility in the Kenyan Cut Flower Industry. *Transactions of the Institute of British Geographers*, 26, pp. 390–406.

ITC (2001) Cocoa. *A Guide to Trade Practices*. Geneva: ITC.

Obi (2000) Outlook for a better creation of value for Sheabutter. Mimeo.

Shukla, V. K. S. (1995) Confectionery Fats. In Hamilton, R. J. (ed.), *Developments in Oils and Fats*. London: Chapman and Hall, pp. 66–94.

UNIFEM (1997) *Le Karite. L'or blanc des africaines*. Dakar: UNIFEM.

Von Maydell, H.-J. (1986) *Trees and Scrubs of the Sahel. Their Characteristics and Uses*. Eschborn: GTZ.

Part II

Commodity chains and cultural connections

4 New geographies of agro-food chains

An analysis of UK quality assurance schemes

Carol Morris and Craig Young

Introduction

Public and political concerns about food are profoundly affecting the operation of agro-food chains. One illustration is the growing emphasis on quality food and the development of 'quality assured' agro-food chains. In the form of quality assurance schemes (QAS), new demands for quality involve a number of organizational developments along the length of agro-food chains, such as modifications to production and processing practices and shifts in the nature of relationships between agro-food actors. Quality assured agro-food chains transform both material and symbolic geographies of agro-food production and consumption. In this chapter we focus on the symbolic dimensions of quality assurance, by which we mean the 'representational' aspects of quality that are produced and consumed by actors within agro-food chains and which enable the material elements of the chain to function. Our approach is influenced by the 'new economic geography' which highlights the importance of the interconnections between material-economic and socio-cultural processes (Lee and Wills, 1997; Thrift and Olds, 1996). Specifically, we are interested to analyse the ways in which quality assured food products are ascribed with distinctive meanings as they move through food supply chains, and to emphasize the highly contested nature of this process. Socially constructed meanings about quality have a number of effects. They enable key actors to exert control over (parts of) the food supply chain in order to maintain and increase their competitive position within that chain; they assist in the process of adding value to food commodities and they facilitate the promotion of the consumption of food products. Drawing upon documentary and social survey data, the chapter considers the ways in which representations of quality are bound up with apparent attempts to reconnect producers and consumers within agro-food chains.

The chapter first considers the importance of quality in agro-food systems and suggests that QAS represent a new sphere of regulation within UK agro-food chains. We then draw upon Cook and Crang's (1996) notion of 'geographical knowledges' in order to analyse the ways in which QAS attempt to imbue food products with meaning. The chapter concludes by assessing the significance of the circulation of representations of quality for UK agro-food systems and producer–consumer relations.

Quality in agro-food systems

It is now widely accepted that there has been increased emphasis placed on the notion of quality within the agro-food system in recent years. Marsden (1997) suggests that the attribution of quality criteria along food supply chains is becoming increasingly ingrained. Acknowledgement of the need to pay attention to quality in agro-food production is also evidenced in frequent statements to this end made by actors within the farming and food industries and reported in the farming press (Morris and Young, 2000).

A variety of economic, political and socio-cultural pressures help to explain the emergence of quality as an important concept within the UK agro-food system. Enhancing quality is seen as an important means of improving the 'competitiveness' of food products. The development of this process in both farming and the food industry is ongoing, but has taken on increased salience in the context of falling farm incomes and an inelastic food market. The production of foods differentiated on the basis of quality marks/labels and quality assurance procedures offers a potential mechanism for guaranteeing outlets, achieving increased sales and premium prices. It also enables particular actors to increase their control over supply chains, thereby improving their competitive position vis à vis others within the same food chain (Marsden *et al.*, 2000). A series of food scares, notably the bovine spongiform encephalopathy (BSE) crisis, also have raised awareness within the farming and food industries about the need for traceability and due diligence over the quality and health status of food.[1] Responsibility for quality and safety also is demanded by national and EU legislation such as the 1990 UK Food Safety Act. Alongside economic and political drivers, socio-cultural factors in food consumption also are influential in the development of the quality food market. Consumer concerns about the nature of food production and in particular its health and safety implications and animal welfare impacts are growing.[2] Increasing demand for quality foods also can be seen to derive from a group of relatively affluent, well-educated consumers who use the purchase of particular types of food to indicate possession of 'cultural capital' (Bell and Valentine, 1997; May, 1996).

Although widely used within discussions of agro-food systems, the notion of quality is often poorly defined and understood. Quality is 'a complex term carrying a vast array of meanings...[referring] to a wide range of criteria and properties' (Watts and Goodman, 1997, 19), and therefore must be conceptualized as a social construct (Ilbery *et al.*, 1999; Marsden and Arce, 1995). This is visible in the commonly used indicators or dimensions of quality within the food industry listed in Table 4.1. Factors such as nutrition, food's sensual attributes and their biological composition may appear to relate more to the physical (and thus 'objective') characteristics of food, but even these are bound up with cultural understandings of their meanings. The appeal of sensual attributes may differ considerably between cultures, and ideas about nutrition have varied over time and are manipulated in different ways in the differentiation of foods (such as the promotion of various non-animal spreads such as margarine). The remaining quality characteristics are more dependent upon their social construction and

Table 4.1 The commonly discussed dimensions of 'food quality' in the food industry

Aspect of food quality	Description
Method of production	'Traditional' methods, welfare/environmentally friendly, socially just.
Place of production	Regionally/locally distinct product.
Traceability	Food has a clearly defined provenance.
Raw materials/content	May relate to perception of the authenticity and naturalness of food.
Safety	Consumer confidence in the safety of production, processing, packaging, labelling, distribution, storage of food.
Nutrition	Food provides a good source of nutrients and meets dietary concerns.
Sensual attributes	The way in which food appeals to the senses, that is, appearance, freshness, texture and flavour, taste, feel and smell.
Functional	The food fulfils the purpose for which it was intended. the
Biological	food supports natural life, for example, in live yoghurt.

Sources: Jacobs (1996), Ilbery and Kneafsey (1997).

contestation – what constitutes safety, is food 'natural' or 'authentic', and what is ethically acceptable in the production of food.

The complex and socially constructed nature of quality also is demonstrated by varying use of the term by different actors within the agro-food system. Ilbery and Kneafsey (1997) have suggested that while producers may regard quality as a marketing opportunity, consumers may relate quality to concerns over food safety or emphasize 'subjective' indicators of quality such as taste, flavour and appearance. Regulatory institutions may be concerned with so-called 'objective' indicators of quality, such as the application of hygiene requirements, although, as Ilbery *et al.* (1999, 5) argue, 'the very objectivity of these indicators is socially constructed and will vary according to political and economic pressures, scientific understanding and cultural contexts'.

A further illustration of the socially constructed nature of quality is in its development and application within two very different spheres of agro-food production: niche and mass food markets. For some commentators quality is necessarily associated with 'alternative' food supply chains which are geographically 'embedded' and developed in opposition to global supply chains associated with mass food markets (Marsden, 1998; Murdoch *et al.*, 2000).[3] For the originators of these alternative and locally oriented supply chains, quality is the antithesis of quantity. In the context of French and Californian wine production, for example, the quality designation 'appellation controllee' is in part contingent upon a wine being produced in limited quantities (Moran, 1993). Similarly, other 'quality' food products are frequently distinguished from mass-produced foods in terms of both place and quantity of production (Ilbery *et al.*, 1999; Marsden, 1998). An example is provided by the EU's 'certificates of special character': PDOs (Protected Designation of Origin); PGIs (Protected Geographical Indication); and TSG (Traditional Speciality Guarantee) (Ilbery *et al.*, 2000). However, quality is not solely a feature of specialist food markets. Through quality assurance procedures, notions of quality are being introduced into the mass food market. This has occurred through nationally and internationally recognized quality management or assurance systems (such as the Hazard Analysis and Critical Control Point System), and via the establishment by multiple retailers of new supply chains based on a particular QAS. Here, quality and quantity coexist, highlighting the ways in which notions of quality can be appropriated in various and competing ways by different actors within the food supply chain.

It is the latter approach to applying quality upon which we concentrate in this chapter. Specifically, attention is focused upon QAS, which have emerged since the late 1980s and arguably are the most explicit illustration of the growing importance of quality in agro-food chains (Morris and Young, 2000; Young and Morris, 1997). The criteria outlined in Table 4.1 suggest that quality does not refer solely to the properties of the food itself but also to the way those properties have been achieved. It also is important to consider the systems and institutional relationships which underpin the production, distribution and retailing of products, of which QAS represents an important example. QAS in the agro-food industry can be understood as a distinct subset of 'quality assurance', which is a widespread phenomenon within both the industrial and service sectors of the economy. QAS entail the application of quality management principles (such as Total Quality Management (TQM)) to agro-food production. These management approaches can be seen as 'transformative philosophies' (Perry *et al.*, 1997, 291) in the sense that they demand a change in the relationships between organizations (greater vertical integration), and in business cultures throughout the supply chain (building trust and sharing knowledge). They are concerned with continuously improving product quality, increasing the efficiency of the supply chain, improving competitive advantage and orientating the supply chain to the interests of the consumer (Hughes and Merton, 1996).

There are numerous QAS in operation and this situation is constantly changing as new schemes are created and existing schemes are amalgamated. Initially associated with fresh meat production, QAS now cover vegetables, fruit, cereals and milk. A variety of agro-food system actors have been responsible for establishing these initiatives, allowing four broad groups of QAS to be identified:

1   farmer and farm industry representative QAS, for example, Farm Assured British Beef and Lamb (FABBL), the Scottish Quality Beef and Lamb Association (SQBLA) and Farm Assured Welsh Livestock (FAWL). A number of these QAS have recently come together under the umbrella of the British Farm Standard 'red tractor';

2   food industry QAS established by abattoirs, food manufacturers and distributors, for example, Anglo-Beef Processors' Total Quality Management System (ABP TQM), and the Birdseye Pea Sourcing Policy;

3   food retailer quality or farm assurance schemes, for example, Sainsbury's Partnership in Produce and Livestock Schemes.
4   QAS established by independent organizations, for example, the Royal Society for the Prevention of Cruelty to Animals Freedom Food's Scheme.

In some cases new organizations and companies have emerged to administer the schemes, some of which involve collaboration between the public and private sectors. For example, Farm Assured British Pigs (FABPIGS) is an independent company owned by the British Pig Association, the National Farmers' Union, the Federation of Fresh Meat Wholesalers and the British Meat Manufacturers Association.

Although the same general principles apply to all schemes, it is acknowledged that a QAS established by a retailer is likely to have slightly different objectives to one set up by the farming industry. Indeed, discussion of farmer-initiated QAS in the farming press frequently centres upon the issue of gaining more control over production standards, and upon marketing food products in the light of the increasing demands of, and regulation by, retailers. However, QAS established 'independently' by the farming industry do not always operate in isolation from retailer QAS: important interlinkages exist between them. For example, under their Partnership in Livestock QAS, the supermarket Sainsbury's sources meat from Anglo Beef Processors, who source livestock only from FABBL accredited farms.

Within the United Kingdom, QAS, specified quality criteria or standards not only apply to production but also to food processing and distribution. On-and off-farm quality criteria therefore can be identified within the scope of individual QAS, and include breed specification, record-keeping to ensure traceability, animal welfare, the application of feed and medicine, environmental management and storage (on-farm standards); and transport, abattoir procedures and hygiene, storage and processing (off-farm standards). There is, however, no common definition of quality and no uniformity in criteria applied within QAS in agro-food production. These differences may be partly accounted for by different goals and levels of influence of actors involved in establishing QAS.

QAS involve a reconfiguration of specific elements of agro-food chains such as on-farm production practices. They also engender changes in the nature of relationships between different parts of these chains, such as the building of trust and cooperation, the formation of producer groups and producer–retailer 'partnerships', and efforts to communicate to consumers how and why food produced within a QAS is of better quality, and how QAS are designed in response to consumers' concerns over agro-food production. Cutting across all of these shifts are the concepts of regulation and reconnection. Although the UK agro-food system has been subject to processes of deregulation during the past two decades, a seemingly contradictory tendency towards re-regulation has been observed in the form of new food legislation and systems of supply chain control operated by powerful food chain actors, notably the retailers (Goodman and Watts, 1997; Marsden *et al.*, 2000). QAS represent a new sphere of regulation within food supply chains, achieved by the application of production and processing standards along the length of food supply chains and a modification in business culture and practices. This re-regulation of food supply systems through the application of quality is in part driven by consumer concerns about food. That some consumers have become more aware of the types of food they eat and food production methods (and, at the same time, producers are being made more aware of the demands of their consumers which they must meet to be competitive) enables the development of quality assurance to be conceptualized as part of a broader process of reconnection between producers and consumers that some commentators suggest is a characteristic feature of contemporary food systems (Buttel, 1998; Marsden, 1998).

The notion of reconnecting producers and consumers raises additional theoretical questions about the role of the consumer in shaping the agro-food system. In both academic and agro-food industry accounts, the consumer tends either to be constructed as knowledgeable (and therefore powerful) or ignorant (and, by implication, manipulated) of the origins of their food. However, Cook *et al.* (1998) argue that there is a need to move away from this 'blunt dichotomy' and to work with more subtle and sophisticated understandings of consumers' relations to systems of provision. Cook *et al.* (1998) identify a 'structural ambivalence' within consumers' relationship with the rest of the food system, involving both an impulse to forget and a need to know about the origins of the food they consume. This suggests a two-way relationship between producers and consumers in which consumers' knowledge of the origins of food are both structured by and help to create relations with food provision. In this chapter we align ourselves with Cook *et al.*'s (1998) positioning of the consumer in relation to other agro-food system actors.

Constructing 'quality' and the 'geographies' of quality assured food

In the United Kingdom, QAS represent a concrete expression of the increasing concern with quality in agro-food systems. Tracing the material geographies of these new approaches to regulating food supply chains would represent one means of understanding their development and operation (e.g. Marsden, 1997; and Morris, 2000 for an account of quality in global agro-food networks). However, as noted above, 'quality' as a concept in agro-food production is socially constructed by actors and analysts and has multiple and contested meanings (see Ilbery and Kneafsey, 1998; Jacob, 1996; Marsden and Arce, 1995; Marsden *et al.*, 1998; Morris and Young, 2000; Murdoch *et al.*, 2000). This suggests that a more discursive analysis of quality is also desirable (Buttel, 1998). This is not to imply that the material and the symbolic are unconnected. As Cook's (1994, 232) analysis of the production and consumption of exotic fruit suggests:

> an examination of the work that goes into the introduction of new – or 'exotic' – fruits to customers suggests that the meanings that companies attempt to ascribe to these fruits play a crucial role in the articulation of

commodity systems. Put differently, just because they are produced and packed in one place and shipped, ripened and delivered fresh to a store in another, it does not necessarily follow that anyone will buy them. In short, there is a *symbiotic relationship* between the *'material'* production of a fruit or vegetable and the *'symbolic'* production of its meanings.

(emphasis added)

Drawing upon these arguments, this section analyses the symbolic geographies of quality assured agro-food production.

The attribution of notions of 'quality' by key food chain actors can be understood as part of a wider process of ascribing social, cultural and symbolic meaning to food commodities. As suggested above, material characteristics of foods are combined with a vast range of societal rituals, norms and understandings of their origins and nature (Bell and Valentine, 1997; Caplan, 1997; Fine *et al.*, 1996; Howes, 1996; May, 1996). Cook and Crang (1996), for example, demonstrate how consumers develop partial understandings of food commodities, which are comprised of various constructed and contested geographical knowledges. The production and distribution of these knowledges form part of the material transformation of food production, distribution and consumption, and are actively created by agro-food chain actors in efforts to gain economic advantage.

Cook and Crang (1996) argue that the construction and use of 'geographical knowledges', or 'symbolic cartographies' (Crang, 1996), involves a number of 'representational processes'. These include 'biographies of food', that is, knowledges of the production and distribution of food; 'origins of food', that is, the construction of geographies of specific places or regions of product origin in which the essential qualities of a product are associated with some essence of place; and 'interest construction', through which actors claim to know and represent other actors in the commodity system. Cook and Crang's (1996) notion of geographical knowledges about food provides a useful framework for understanding the construction and contestation of quality in agro-food production. We explore the production and deployment of each of the geographical knowledges through analysis of technical documentation, packaging, promotional material and media coverage of specific QAS, and interview data with various types of actors in the agro-food chain. Consideration of the use of standards in ascribing quality provides an understanding of 'biographies of food'. We consider issues surrounding the 'origins of food' through an analysis of how QAS create and promote particular versions of the places in which quality foods are produced. How key QAS actors represent consumers comprises our analysis of 'interest construction'.

Biographies of quality food: the construction of standards

Within the United Kingdom, QAS 'quality' is partly defined by the development of particular sets of standards. These are devised by the organizations and companies that establish and operate QAS and are designed and employed to control the aspects of food production, processing and distribution to which the QAS applies. The ascription of quality to food is intended to be achieved by developing a seemingly objective set of criteria. However, this is a variable and contested process, and is deeply embedded in social, economic and political relations.

Although attempts are currently being made to develop 'base line' standards for QAS initiatives, there is considerable variation in the nature and use of standards. Tables 4.2 and 4.3 present 'actual' quality standards with which producers must comply. We focus upon two QAS that apply to fresh meat: the FABBL scheme and the Avonmore Waterford scheme (Ireland). Although these standards are highly detailed, there is variation in this detail which attempts to 'build in' quality to the same type of product. While these details seek to apply quality standards to production processes rather than to the food commodities themselves, their socially constructed nature is still apparent at this stage. Examining the ways in which the application of standards in production systems are ascribed to food products in attempts to construct an image of quality reinforces this point. The complexity of QAS requirements, outlined in great detail in scheme manuals, is considerably simplified in representations of the process for the consumer. Promotional material for the Co-op's Farm Selected British Beef, for example, states that 'selected farms are independently monitored to ensure that only the highest standards of farming practice are used. Traceability and monitoring of on-farm standards is assured.' Under 'food safety', the promise is that 'strict codes of practice ensure that only high quality feeds are used'.[4] SQBLA's promotional material for Speciality Selected Scotch Beef gives details of farm assurance and quality assurance in the production and distribution of their products. The beef is claimed to be 'safe and wholesome' and can be 'traced right back to its farm of origin'. Independent inspectors 'check...farms adhere to

Table 4.2 Quality standards in the FABBL scheme

Standards
- Origin of livestock handled, feed composition and storage, medicines and veterinary treatments used, husbandry and welfare, housing and handling facilities, movement records and medicine books, identification and marking, transport to national outlets.
- National set of standards for production and transport of British beef and lamb. Designed to meet retailer demands at farm production and transport stages, and to comply with all current codes of recommendations and legal requirements.
- Full list of FABBL on-farm standards in control manual.

Definition of quality
Refers mainly to quality of finished product and immediate environment of farm animals. Promotional material makes reference to 'wholesomeness' of meat product from a consumer point of view, and a concern for the welfare of animals on and off farm.

Source: Authors.

Table 4.3 Quality standards in the Avonmore Waterford Farm Assured Beef (AFAB) system

---

Standards
- Standards imposed by the system include: pollution control; control over the volume of medicines and veterinary products applied.
- Vet visits must be recorded, quality of animal feed must be within the control limits of the system, ingredients used for feeding livestock must be tested and meet the requirements, silage quality must be of a certain standard.
- Controls over the use of pesticides and chemical fertilizers are maintained.
- Standards are set for the space required by animals over the winter.
- A code of practice for transportation of animals is followed closely, with particular emphasis on the loading and unloading procedures.
- There is a procurement policy which limits the availability of a producer to purchase young stock other than from an accredited producer of calves.
- Other standards specifically relating to the animals include: number of hours spent under artificial lighting, water availability at the processors and length of time held in liarage.

Definition of quality
- Precise definitions of quality are indicated on pages 17–33 of the AFAB operations manual, but in particular, animal welfare and environmental issues are considered to be important.
- Quality in relation to the finished animal prior to arrival at the factory is not defined, other than it has to have a minimum grade standard of R4L.

Source: Authors.

codes of practice governing animal health, welfare and feeding'. At the meat processing stage, 'standards are rigorously enforced'.

These examples demonstrate that the promotion of quality food products does attempt to respond to consumer concerns by highlighting the application of quality standards to food production and distribution. There is some attempt deliberately to 're-connect' the consumer to the processes of food production, manufacturing and distribution. However, the information supplied is highly simplified and consists largely of rather bland assurances on vague points (e.g. meat is 'safe and wholesome'). In addition, this portrayal of the application of standards is not really telling the consumer anything new. Much discussion about standards refers to practices required by law which should already be undertaken by food producers.

The introduction of quality standards is a part of the process of constructing particular biographies of food. However, this is not an uncontested process. In 1998, farmers made political representation through the Federation of Small Businesses to the UK Office of Fair Trading and the European Commission to mount a legal challenge to QAS. Resistance is based on perceptions that farmers are being forced to join QAS but are receiving few benefits and incurring considerable costs in doing so (Morris and Young, 2000). This resistance is fuelled by farmers' attitudes to processes of ascribing attributes of quality to their food products in the construction of food biographies. On the one hand, political challenges are based on whether the processes of ascription are meaningful in the eyes of producers. On the other, there is opposition to a process of non-ascription of quality to food produced by farmers who are not members of QAS. Farmers do not resist 'quality' labels *per se*, but rather question whether processes of ascription or non-ascription are economically and politically acceptable. The construction of 'quality' by powerful actors within (and often further up) the agro-food chain is thus a politically contested process. 'Quality' is not a simple economic strategy, but is socially constructed and politically resisted.

The origin of food: quality assured food and the construction of 'place'

The process of ascribing quality to food in QAS also draws upon the 'discursive construction of various imaginary geographies' (Cook and Crang, 1996, 134). QAS bases distinctions upon 'natural' characteristics such as 'pure' soil, water and air and upon the 'traditional' food culture of regions or localities in which food is produced. A number of examples of QAS for meat and cereals, fruit and vegetables are used to illustrate these points. We identify both farm industry and retailer QAS, although similar themes are evident in both.

SQBLA expanded its promotion from within the meat trade to a £500,000 national television advertising campaign in 1998. Their newly appointed marketing manager stressed that 'Brand awareness is a crucial factor in ensuring that Scotch beef and lamb expand their market share thereby increasing value which will eventually filter back to the primary producer' (SQBLA Newsletter, Spring, 1998). The information pack folder for SQBLA is fronted by the message of 'total quality assurance from farm to consumer', stressing the importance of the application of quality along the agro-food supply chain. The cover depicts a stereotypical image of the Scottish Highland landscape, comprising rushes, a loch and mountains. Livestock and grazing land are absent. The SQBLA newsletter also avoids portraying livestock or grazing land but uses imagery of mountains and heritage to suggest the Highland landscape. A leaflet promoting Specially Selected Scotch Beef again portrays a Scottish landscape, and the text emphasizes Scotland is 'a land famed for its mountains, rivers and rich pastures'. On the reverse, cows in improved grazing land are represented, but the livestock are so small as to be unidentifiable. Similar depictions are evident on promotional leaflets for Scottish Quality Trout and SQBLA Scotch Lamb, where quality assured lamb is described as 'the natural choice', from the 'clean, natural environment' of 'lush Scottish pastures'. Likewise, the Welsh Beef Mark is aimed at consumers who appreciate the 'clean natural environment' in which Welsh beef is raised (*Farmers Weekly*, December 1996).

Similar themes also can be seen in retailer based QAS for meat. For example, the Co-op's Farm Selected British Beef promotional leaflet does not feature any animals. Rather, a generic rustic agricultural landscape is presented, with graphics featuring cereal production, and a farm gate without a fence. Marketing materials seek to promote 'traditional' forms of agriculture, involving less intensive farming methods. A somewhat different portrayal of quality assured meat production is depicted on promotional material for Sainsbury's Partnership in Livestock Scheme (see Plate 4.1). Although cattle are described as providing 'Nature's finest beef' the leaflet does incorporate elements of productive landscapes alongside illustrations of features such as hedgerows, water courses and traditionally

Plate 4.1 Promotional leaflet for the Sainsbury's Partnership in Livestock programme.

Source: Sainsbury's Supermarkets Ltd.

managed pasture that are commonly interpreted as 'natural'. The construction of images of quality is a complex process involving competing representations. Yet another set of associations with place and landscape appear in the 'Cornish King' label for potato marketing. As a buyer suggested, the label 'exploits the concept of Cornwall as a place of pleasant holiday memories. This gave the label a huge lead… Cornwall has a perception advantage – cliff tops and beaches and good things and holidays. Let's capitalise on it' (Univeg manager, *Farmers Weekly*, June 1996).

A number of common themes within this set of representational processes can be identified. First, animals from which quality assured meat products are derived rarely are represented. There is an absence of animals depicted in productive locations (e.g. fields, farm buildings), a general absence of carcasses, whole dead animals and the processing of animals. One exception is the Scottish Quality Trout scheme, the promotional materials for which do depict whole dead fish and prepared fish. A camouflaging of the relationship between meat and animals may be partly explained by market research which has found that consumers do not want to make links between animals and the meat they consume (*Farmers Weekly*, November 1995). Food consumers often establish 'cognitive thresholds' (Cook *et al.*, 1998) in order to manage the complexity of the food provision system, and beyond which they do not routinely venture.

Second, productive agricultural landscapes associated with meat and fruit and vegetable production often are absent. Images seek to invoke a 'natural' environment, rather than intensively managed pastureland or fields of vegetable crops. It is rare to encounter images of any productive processes such as handling, transporting or slaughtering animals, ploughing, applying chemical inputs or disposal of waste. Exceptions exist, but primarily are found in QAS documentation aimed at persuading producers to become scheme members (such as the Assured Combinable Crops Scheme), rather than in promotional material aimed at consumers.

A third common theme involves representations of wildness, purity or cleanliness of the locality or region from which the food originates. Thus 'natural' landscapes or environments are used to ascribe meaning to quality assured foods. Quality is seen to derive from the production of food in wild, natural and pure or clean environments. The quality of air, water and the environment in general is stressed either implicitly in the text of promotional material or explicitly through the use of imagery. This is often linked to representations that associate environmental concepts with particular places such as Wales and Scotland.[5]

Finally, there is an implicit or explicit implication that food production methods associated with QAS are environmentally beneficial, or at least 'care' for the environment in some way. The suggestion is that QAS directly respond to consumer concerns about the environmental impact of food production. However as research into the environmental standards has revealed, the extent to which this is actually the case is questionable (Morris, 2000).

Our analysis raises questions about the degree to which the development of quality food markets is part of a genuine process of re-connecting producers and consumers. Insofar as they seek to allay consumer fears over food production methods, QAS at least ostensibly aim to 're-connect' the consumer to producers and commodities. However, efforts to increase consumers' 'geographical knowledges' about the origins of the food appear primarily to involve a 're-imaging' of food commodities to allay fears, through a representation of quality assured foods as 'natural'. The consumer, it appears, receives a new, more acceptable image of agro-food production, but at the same time the realities of food production and the potential benefits of QAS are obscured. QAS could therefore be understood as contributing to consumers' 'structural ambivalence' about the food system. While apparently providing more knowledge of the origins of food to consumers, QAS conceal or provide partial 'truths' about origins in order that consumers are not overwhelmed by information. Alternatively, QAS could be conceptualized as one of the tools used by consumers in constructing cognitive thresholds about the food system. As Cook *et al.* (1998, 164) argue we should not be working towards the 'all-knowing' consumer, 'instead, the issue [...] is more one of the institutions and sites in which consumers can place responsibility for knowledge of food provision'. QAS could be interpreted as an example of such institutions and sites.

Interest construction: representing the consumer

The creation of representations of quality food's biographies and origins also is linked to 'interest constructions' (Cook and Crang, 1996). Through food quality schemes, producers and retailers not only ascribe meaning to the food itself but also construct versions of 'the consumer' and of 'consumer interests' (Marsden *et al.*, 1998, 468). One aspect of ascribing 'quality' to food thus involves representational processes that create an image for food which communicates the message that consumer concerns have driven its production. This process also necessarily involves the construction of the consumer by key actors in QAS and can be illustrated through an examination of discourses within a variety of UK print media (Morris and Young, 2000). Frequent statements are made about the need to respond to consumer concerns: 'the consumer is king and is demanding specific assurances on quality, safety and welfare' (*Farmers' Weekly*, June 1997).

However, consumers are represented in a particular way, and there has been a marked evolution in these representations since the early 1990s. Previously, there was little acknowledgement that consumer demands were of direct significance to farmers' decision-making. Now there are frequent references to the 'powerful', 'demanding' and 'discerning' consumer, whose increased awareness and concerns about food production techniques, related demands for quality, safe and nutritious food are seen to drive changes in production and food marketing. For example, according to the head of Genus Management, consumers are 'demanding assurances that their milk has been produced to top quality standards, under good welfare conditions and that it can be traced back to an individual farm' (*Farmers' Weekly*, May 1996). Consumers are portrayed as knowing what they want in food production, with QAS being a response to 'today's highly selective consumer demands' (ADAS representative, reported in *Farmers' Weekly*, October 1995). It is reported that 'consumers are increasingly interested in meat quality, hygiene and welfare...and are prepared to pay more if they know it is of assured quality or produced in a welfare-friendly way...All the signs suggest the current interest in food and how it is produced will become the norm, rather than the exception' (*Farmers' Weekly*, editorial statement, August 1995). Further, 'consumer confidence is relying more and more on products sold into the food industry being quality assured and traceable' (Farm management consulted reported in *Farmers' Weekly*, May 1996).

Representations of the consumer in the farming media appear to be constructed on the basis of limited evidence about consumer preferences, concerns and behaviour. Although the major retailers conduct extensive consumer research, much that is written about the 'demanding consumer' does not make any reference to this type of empirical work. For example, the SQBLA scheme was developed through consultation with the farming community at meetings and road shows. Farmers developed key standards for housing and handling of animals, feed storage and the use of medicines. Scheme administrators felt that 'no surveying of retailers and consumers was considered necessary' (Telephone interview, 1998). Nevertheless, documentation promotes beef produced in an 'unbroken chain of quality', incorporating discerning butchers, and 'Scotch lamb – a quality product that clearly meets the needs of discerning customers today' (SQBLA, Specially Selected Scotch Beef and Lamb promotion leaflets, n.d.).

However, the transmission of culturally constructed images is by no means a linear process. Consumers do not receive a promoted image passively and uncritically, but rather attitudes, values and perceptions are bound up in an active process of consumption which extends beyond the purchase of a product (Cook and Crang, 1996; Cook *et al.*, 1998; Crang, 1996). Attention needs to be focused on the critiquing processes of consumers and their active knowledge constructions (Cook and Crang, 1996).

Our own survey work reveals a degree of consumer confusion about the meaning of quality associated with fresh meat products.[6] Responses to an open-ended question about understanding of quality assurance in meat production are summarized in Table 4.4. Thirty-six percent of consumers had a good understanding of quality assured fresh meat, but 64 per cent equated quality with a range of issues, mainly related to food safety, freshness or being of a better standard than non-quality assured food. Other consumers' notions of quality included: attention paid to animal welfare; the regulation of meat producers; an understanding of the regulations relating to cattle and BSE; safe use of fertilizers; and 'telling the truth about products'. Consumers constructed fluid and occasionally contradictory ideas about 'quality' in agro-food production. They also expressed opposition to notions of 'quality' in suggestions that quality marks are merely marketing tools used to justify higher prices.

Nonetheless, 74 per cent of consumers surveyed stated that they would have more confidence in a fresh meat product if it carried a quality assurance label. For the majority, such responses rested upon understandings that labelling indicated

Table 4.4 Consumers' understandings of 'quality assurance'

Meaning of quality assurance	% of consumers
Product has passed certain standards and can thus be labelled 'quality assured'	36
The best quality food you can buy	12
The product is fresh	12
Safe for human consumption	11
Free from harmful bacteria or disease	10
Safety standards have been ensured in food production	8
Animal welfare has been assured	4
A marketing tool to sell goods at higher prices	2
Other	1
Don't know	4

Source: Authors.

Note
n = 190.

Table 4.5 Consumer confidence in quality assured meat

Response to question: would the consumer have more confidence in a meat product if it was labelled 'quality assured'?	% of consumers
Yes, because...	73
Quality standards have been adhered to	32
The produce has been 'checked'	9
The consumer can have confidence in the product	7
Production processes have been regulated by a responsible body	4
The customer can have 'peace of mind'	3
Other	18
No, because...	27
The label doesn't mean anything/don't believe labelling/ there are too many labels	19
It's just a marketing tool	1
Other	1
Don't know	6

Source: Authors.

Note
n = 190.

the achievement of certain standards by fresh meat producers (see Table 4.5). The next most important explanations included a general trust in quality labelled food, and a belief that meaningful regulation of food production had been undertaken by a responsible body. Other factors used in constructions of quality food were the importance of 'Buying British', safety, animal welfare and traceability. However, 26 per cent resisted the notion that a quality-assured labelled product would inspire consumer confidence about the production and distribution of food. The majority of these consumers expressed mistrust or cynicism about the meaning of such labelling and a small proportion regarded it simply as a marketing ploy.

Consumers do not simply 'read off' meanings of quality which QAS seek to create and promote. Confusion surrounding the term suggests not only a reduction in its marketing value but even the creation of direct opposition to its use and meaning. Such a situation reflects broader arguments about the lack of connection between advertising campaigns and consumer tastes and actions (Miller, 1997). Representations of the consumer by both retailers and in the trade press reproduce assumptions about consumer attitudes and understandings of food in order to legitimate the development and operation of QAS. These representations are used to support the introduction of notions of quality into other parts of agro-food supply chains, particularly food producers. A caring and responsive image is then presented to consumers, which suggests that concerns about quality and safety are being addressed by the food and farming industries.

Conclusions

This chapter has sought to reveal the complex and contested manner in which meanings of 'quality' mediate the operation of agro-food chains. The construction of 'quality' is associated with new material geographies of agro-food production but is also used within the agro-food industry in representational strategies aimed at 'reconnecting' the consumer and the producer. The introduction of notions of quality is widely held within the industry to be an important mechanism for allaying consumer concerns about production and therefore stimulating consumption. Not only have we sought to indicate the socially constructed nature of 'quality' but also we have attempted to foreground the extent to which diverse sets of meanings circulate amongst different actors in the agro-food chain.

In particular, it is questionable whether attempts to re-image agro-food production in QAS achieve a reconnection of producers and consumers. There is, perhaps, a partial reconnection of the consumer with food production processes through attempts to inform them about the welfare, environmental and safety issues of the food they consume. However, the diverse ways in which such meanings are constructed and circulated, combined with the complex ways in which consumers 'read' such meanings as a part of the process of consumption, raises questions about how meaningful marketed notions of quality are to consumers. Our analysis of QAS imagery suggests that this 're-imaging' may in fact serve to further obscure the realities of production and distribution. Where consumers are more 'discerning' or demanding of information it may fail to inform them, or even produce a cynical resistance to the introduction of notions of quality.

Given declining incomes within UK agriculture and ongoing struggles which emerged from a serious foot and mouth disease outbreak in 2001, the issue of quality is likely to become more significant for the agro-food industry. For a limited number of producers, alternative strategies incorporating the marketing of quality imagery based on 'localness' and regional distinctiveness may offer a way forward. For the bulk of the industry, however, production may become even more intensified and competitive – although much depends upon future changes to the European subsidy regime within which agro-food production operates. In both cases the issue of quality is likely to remain central in efforts to reassure consumers about agro-food products. Complex connections between the material and symbolic aspects of agro-food production and consumption when represented will continue to shape the circulation of commodities.

Notes

1 At the time of writing, the United Kingdom experienced its worst out-break of foot and mouth disease which threatened to devastate the livestock industry. Although distinct from food scares such as the salmonella and BSE crises – both of which had a more direct impact on human health – it nonetheless has raised similar questions about the operation of meat supply chains.
2 For a more extensive account of the relationship between food consumers, food products, environment, animals and food producers, see Whatmore (1997).
3 Actor network theory would reject such a binary distinction between local/ alternative and global/mass food supply chains (Whatmore and Thorne, 1997). However, it is apparent that agro-food chain actors themselves actively create distinctions in their attempts to define and use quality in different ways.
4 There is a clear reference to the association between BSE and animal feeds.
5 For a discussion of associating quality and regional imagery in the 'West Country' see Kneafsey and Ilbery (2001).
6 The consumer survey was conducted as exploratory research aimed at an initial assessment of understandings of quality in agro-food production and of QAS in particular. Every second consumer leaving a range of retail outlets was approached to complete a questionnaire. The survey was conducted at various days and times throughout the week. Retail outlets were chosen to represent a range of retail types, from large supermarket chains, through smaller and local supermarkets serving the cheaper end of the market, to small 'high street' retailers such as 'family' butchers. Chosen outlets were located in a number of north-west towns and cities including Southport, Leyland, Chorley, Preston, Eccleston, Clayton, Brook, and Didsbury and Fallowfield in Manchester. 190 individuals were surveyed, representing different income levels, marital status, number of dependents, age and shopping behaviour. The final sample contained more women than men (63 and 37 per cent of respondents respectively) but this reflects the fact that women continue to assume responsibility within households for food shopping (Warde, 1997).

References

Bell, D. and Valentine, G. (1997) *Consuming Geographies: We are Where We Eat.* London: Routledge.

Buttel, F. H. (1998) Nature's place in the technological transformation of agriculture: some reflections on the recombinant BST controversy in the USA. *Environment and Planning A*, 30, pp. 1151–63.

Caplan, P. (ed.) (1997) *Food, Health and Identity.* London: Routledge.

Cook, I. (1994) New fruits and vanity: symbolic production in the global food economy. In Bonanno, A., Busch, L., Friedland, W., Gouveia, L. and Mingione, E. (eds), *From Columbus to ConAgra: The Globalisation of Agriculture and Food*. Kansas: University Press of Kansas.

Cook, I. and Crang, P. (1996) The world on a plate: culinary culture, displacement and geographical knowledges. *Journal of Material Culture*, 1, pp. 131–53.

Cook, I., Crang, P. and Thorpe, M. (1998) Biographies and geographies: consumer understandings of the origins of foods. *British Food Journal*, 100, pp. 162–7.

Crang, P. (1996) Displacement, consumption and identity. *Environment and Planning A*, 28, pp. 47–67.

Fine, B., Heasman, M. and Wright, J. (1996) *Consumption in the Age of Affluence: The World of Food*. London: Routledge.

Goodman, D. and Watts, M. (eds) (1997) *Globalising Food: Agrarian Questions and Global Restructuring*. London: Routledge.

Howes, D. (ed.) (1996) *Cross-Cultural Consumption: Global Markets, Local Realities*. London: Routledge.

Hughes, D. and Merton, I. (1996) Partnership in produce: the J. Sainsbury approach to managing the fresh produce supply chain. *Supply Chain Management: An International Journal*, 1, pp. 4–6.

Ilbery, B. and Kneafsey, M. (1997) Regional images and the promotion of quality products and services in the lagging regions of the European Union. Paper presented at the third Anglo-French Rural Geography Symposium, Nantes, September 1997.

Ilbery, B. and Kneafsey, M. (1998) Product and place: promoting quality products and services in the lagging rural regions of the EU. *European Journal of Urban and Regional Studies*, 5, pp. 329–41.

Ilbery, B., Kneafsey, M. and Bamford, M. (2000) Protecting and promoting regional speciality food and drink products in the European Union. *Outlook on Agriculture*, 29, pp. 31–7.

Ilbery, B., Kneafsey, M., Bowler, I. and Clark, G. (1999) Quality products and services in the lagging rural regions of the European Union: a producer perspective. Paper presented at the International Rural Geography Symposium, Nova Scotia, July 1999.

Jacobs, M. (1996) The role of safety in the quest for food quality. *Environmental Policy and Practice*, 5, pp. 91–5.

Kneafsey, M. and Ilbery, B. (2001) Regional images and the promotion of speciality food and drink in the west country. *Geography*, 86, pp. 131–40.

Lee, R. and Wills, J. (1997) *Geographies of Economies*. London: Arnold.

Marsden, T. (1997) Creating space for food: the distinctiveness of recent agrarian development. In Goodman, D. and Watts, M. (eds), *Globalising Food: Agrarian Questions and Global Restructuring*. London: Routledge.

Marsden, T. (1998) New rural territories: regulating the differentiated rural spaces. *Journal of Rural Studies*, 14, pp. 107–17.

Marsden, T. and Arce, A. (1995) Constructing quality: emerging food networks in the rural transition. *Environment and Planning A*, 27, pp. 1261–79.

Marsden, T., Flynn, A. and Harrison, M. (1998) Creating competitive space: exploring the social and political maintenance of retail power. *Environment and Planning A*, 30, pp. 481–98.

Marsden, T., Flynn, A. and Harrison, M. (2000) *Consuming Interests: The Social Provision of Foods*. London: UCL Press.

May, J. (1996) A little taste of something more exotic: the imaginative geographies of everyday life. *Geography*, 81, pp. 57–64.

Miller, D. (1997) *Capitalism: An Ethnographic Approach*. Oxford: Berg.

Moran, W. (1993) The wine appellation as territory in France and California. *Annals of the Association of American Geographers*, 83, pp. 694–717.

Morris, C. (2000) Quality assurance schemes: a new way of delivering environmental benefits in food production? *Journal of Environmental Planning and Management*, 43, pp. 433–48.

Morris, C. and Young, C. (2000) 'Seed to shelf', 'Teat to table', 'Barley to beer' and 'Womb to tomb': discourses of food quality and quality assurance schemes in the UK. *Journal of Rural Studies*, 16, pp. 3–15.

Murdoch, J., Marsden, T. and Banks, J. (2000) Quality, nature and embeddedness: some theoretical considerations in the context of the food sector. *Economic Geography*, 76, pp. 107–25.

Perry, M., LeHeron, R., Hayward, D. J. and Cooper, I. (1997) Growing discipline through total quality management in a New Zealand horticulture region. *Journal of Rural Studies*, 13, pp. 289–304.

Thrift, N. and Olds, K. (1996) Refiguring the economic in economic geography. *Progress in Human Geography*, 20, pp. 311–37.

Warde, A. (1997) *Consumption, Food and Taste*. London: Sage.

Watts, M. and Goodman, D. (1997) Global appetite, local metabolism: nature, culture and industry in fin-de-siecle agro-food systems. In Goodman, D. and Watts, M. (eds), *Globalising Food: Agrarian Questions and Global Restructuring*. London: Routledge.

Whatmore, S. (1997) Dissecting the autonomous self: hybrid cartographies for a relational ethics. *Environment and Planning D: Society and Space*, 15, pp. 37–53.

Whatmore, S. and Thorne, L. (1997) Nourishing networks: alternative geographies of food. In Goodman, D. and Watts, M. (eds), *Globalising Food: Agrarian Questions and Global Restructuring*. London: Routledge.

Young, C. and Morris, C. (1997) Intra-industry linkages in the agro-food chain: new forces for regulating aspects of food production? Paper presented at the RGSG session 'Regulating Rural Spaces', at the Annual Conference of the RGS-IBG, University of Exeter, January 1997.

5 Culinary networks and cultural connections

A conventions perspective

Jonathan Murdoch and Mara Miele

Introduction

In *Sociology beyond societies* Urry (2000) outlines a new global configuration of commodity chains and production networks. He describes innovative interminglings of social, technological and natural phenomena within chains and networks and explains how 'human powers increasingly derive from the complex interconnections of humans with material objects, including signs, machines, technologies, texts, physical environments, animals, plants, and waste products' (Urry, 2000, 14). Globalization refers to the increasing reach of economic networks as they align heterogeneous arrays of interrelated entities in ways that ensure their products and services are distributed widely.

Urry (2000) draws attention to the heterogeneous composition of global networks in order to indicate how traditional forms of stability associated with discrete societies have given way to a world of flows in which goods, images, peoples, technologies and other artefacts traverse long-established boundaries and borders. He takes this development as a starting point from which to argue that sociology – which has long taken social structures within nation-states as primary objects of analysis – should engage more wholeheartedly with *mobility*. Given the heterogeneous nature of global flows, Urry (2000) claims that sociology must study more than the 'social': it must focus on 'mixtures' and on zones where differing entities and artefacts 'exchange properties' (Latour, 1999) as longer and more complex networks are constructed. Urry (2000) believes a new form of sociological analysis is required which can think 'beyond society' and recast its subject matter into the form of networks, commodity chains, fluid social spaces and global institutional forms.

This prognosis for sociology clearly has implications for geographical work on commodity chains, not least the need to examine the full range of interconnections between networks and spaces in a context of growing mobility (see also Castells, 1996). The recognition that network 'heterogeneity' provides a useful starting point for investigating how commodity chains interact with spatially distributed resources is becoming increasingly evident within economic geography (Amin and Thrift, 1995; Dickens *et al.*, 2001; Murdoch, 1995). In this chapter, we develop this general approach further by considering different ways in which network relations might 'coalesce' with given spatial formations. In so doing, we introduce a note of caution into Urry's argument: we propose that while networks might appear to be increasingly mobile and increasingly disconnected from given spaces, they also act to 'condense' space and time. In other words, socio-economic networks and commodity chains emerge from spatially-locatable values, symbols, products, practices and entities; and they serve to reflect and refine these phenomena within network forms (see also Bridge, 1997; Kirsch, 1995; Murdoch, 2000). We believe this process of 'condensation' – which ensures a continuing interrelationship between networks and territorially distributed resources and entities – requires further analytical attention.

In looking more closely at how networks 'condense' space we concentrate on the role played by cultural frameworks in linking chains and territories. It has long been noted that the globalization of economic networks coexists with relatively stable cultural formations (Hannerz, 1992), so that networks both of standardization and localized variability are invariably present in contemporary spatial configurations (Murdoch, 1998). We therefore pay particular attention to the ways in which cultural frameworks interact with heterogeneous materials so that network and spatial formation reach some kind of (albeit temporary) accommodation with one another.

In connecting heterogeneous networks to cultural formations, we adopt a conventions perspective. This approach seeks to identify how social actors combine resources (including both things and people) in line with particular 'orders of worth' (Boltanski and Thevenot, 1991; Lamont and Thevenot, 2000). By focusing on the forms of calculation that lie behind 'orders of worth', the theory aims to understand how cultural norms and heterogeneous networks mutually consolidate one another. In other words, culture is encoded within complex arrangements of diverse materials, while diverse arrangements of materials are constructed according to cultural norms.

We apply conventions theory to an economic arena in which connections between heterogeneously composed networks and cultural formations remain relatively strong: the food sector. Although this sector has undoubtedly been 'globalized' in the sense that economic, social, technological and ecological aspects are promiscuously intertwined across vast spaces (Bonnano *et al.*, 1994; Goodman, 1999; Goodman and Watts, 1997), regularities and stabilities in food production practices remain. Mobile networks of globalization continue to coexist with discretely spatialized production forms (cf. Ilbery and Kneafsey, 1998; Whatmore, 1994). It is thus likely that differential relationships between commodity chains and cultural formations will be evident within the food sector.

We also use the conventions approach to investigate how the composition of food commodity chains is linked to new patterns of food consumption (see Dixon, 1999; Lupton, 1996; Warde and Martens, 2000). As global agro-food networks generate an abundance of food commodities (especially in advanced capitalist countries), consumers are provided with a wider range of food choices (Montanari, 1994). It seems likely that increasingly sophisticated consumers will utilize a range of cultural repertoires in order to choose between a wider array of food products (Cook *et al.*, 2000; Miele, 2001). We can therefore speculate that cultural forms of evaluation will play an expanding role in shaping food commodity chains (Cook and Crang, 1996; Lowe and Wrigley, 1996; Miller *et al.*, 1998).

We suggest that an analysis of heterogeneous networks from the standpoint of conventions theory may make it possible to consider how consumer conventions (linked to quality and value) come to be assessed against producer conventions (linked to efficiency and cost). However, in pursuing this approach it also is necessary to follow the commodity chain methodology by considering how 'intermediate' actors such as manufacturers, retailers, restaurant chains, caterers and food movements also promote given mixtures of conventions. By combining commodity chain and conventions perspectives we view food commodity networks as complex processes in which different conventions are 'traded' against each other (Wilkinson, 1997).

In order to evaluate how conventions in food commodity chains come to be negotiated and aligned by a range of 'intermediate' actors we have chosen to focus in this chapter upon the 'culinary network', a concept that refers not to a single commodity (as in commodity chain analysis), but to the array of materials, actors and institutions that comprise *cuisines*. Cuisines are stable arrangements of foods, ingredients, practices, and tastes and comprise cultural and productive 'worlds of food'. The close connection between food and culture within cuisines makes the 'culinary network' an especially appropriate arena in which to link production and consumption using the conventions approach.

To facilitate an investigation of differing mixtures of actors and conventions in culinary networks we have selected case studies that highlight very different 'worlds of food'. The first is perhaps the archetypal example of 'globalization' in the food sector – McDonald's. Although this network is largely based upon standardized network relations, it also incorporates heterogeneous resources harnessed in line with specific cultural conventions. In particular, this assemblage reflects a set of cultural norms linked to the time and place of the restaurant chain's establishment: the post-war suburban United States. We examine how McDonald's acts to refine and disseminate cultural norms and we consider how a mix of conventions has become encoded in the heterogeneous materials that facilitate the flow of a 'McDonaldized' cuisine.

Having outlined the interaction between standardized network and cultural convention in the case of a global chain, we turn to examine a second case study which displays a markedly contrasting set of connections between network and culture. The Slow Food movement has arisen explicitly to challenge the 'McDonaldization' of cuisine and aims to reassert the values of foods in diverse spatial contexts. It emphasizes that cuisines should reflect localized cultural norms and practices. For Slow Food, cuisine variation goes hand in hand with cultural variation. The movement thus disseminates a range of conventions associated with local food cultures perhaps most evident in its country of origin: Italy.

In contrasting the two cases, we show that each network reflects the general repertoires of evaluation found within two main cultures of food, one 'fast' and the other 'slow'. While the networks might be seen to reflect these cultures in terms of the conventions employed, a recursive relationship between network and culture is evident; that is, the way a network condenses conventions has an impact on the culture of food. We therefore propose that networks and their (cultural) contexts mutually constitute one another. Before presenting this analysis we outline recent theories of food commodity chains in order to show how material complexity and multiple conventions have become of increasing concern to those studying the geography of food commodity networks.

Chains, networks and conventions in the food sector

The investigation of commodity chains or networks in the food sector has strong theoretical roots. The first examples of agro-food commodity chain analysis appeared during an early round of Marxist theorizing. The political economy of food chains identified an increasingly rapid destruction of traditional agricultural production forms (e.g. family farms) as the imposition of capitalist relations fuels a process of industrialization (de Janvry, 1981). This industrialization process appears to be 'disembedding' food production from its pre-existing ('pre-industrial') economic, social and spatial connections. Work conducted in the United States by Friedland *et al.* (1981), for example, discerns differential rates of capitalist penetration in the agro-food sector but concludes that the process is well advanced across the food sector as a whole. Within each commodity chain, differing mixtures of technical, natural and economic resources are integrated so that distinctive industrial structures (of which agriculture is a diminishing part) are evident. The notion of 'commodity chain' is adopted because it shows how different commodity sectors are organized and highlights the complex sets of relationships invoked within each organizational segment.

The political economy of commodity chains was tailored to the sets of relations that are typically constructed around different agro-food commodities. Friedland (1984) summarizes the research foci of the early studies as: the labour process; grower and labour organizations; the organization and application of science and technology; and distribution and marketing. As this list indicates, commodity system studies dealt largely with the economic and social dimensions of industrialization (see Buttel *et al.*, 1990). Friedland (2001, 84) recently has admitted that analysts frequently took as their main concern 'agricultural mechanisation and its social consequences'. They therefore tended to link network and spatial context primarily through an analysis of the industrial reconfiguration of space and how the industrial rationalization of the chains reconfigures production relations at the local level.

More recently food sector analysts have drawn attention to another aspect of commodity chain activity: the environmental or natural components that often are central to food chain construction both in terms of production (e.g. seasonality, perishability, pollution) and consumption (e.g. quality, health, safety). In early work, Friedland *et al.* (1981) noted that the specific nature of particular agro-food commodity chains often is determined at least in part by the natural properties of the commodity itself (e.g. the perishability of lettuce and tomatoes). This insight is developed by Goodman *et al.* (1987) who suggest that the consolidation of capitalist enterprises in the food sector involves the replacement and substitution of natural processes as part of an effort to remove biological constraints from the production process (see also Goodman and Redclift, 1991). Goodman *et al.* (1987) also argue that an expansion and lengthening of food networks tends to result from the progressive industrialization of food so that food products come to be transported over longer and longer distances. This lengthening of food chains increases their socio-technical complexity and leads to the emergence of global commodity chains.

Despite the wealth of valuable insights generated in these studies, commodity chain analysts tend to regard the development of specific chains as conforming to a simplified set of industrial rationalities. According to Busch and Juska (1997) it is usually assumed that the chains express multinationals' 'will to power' (cf. Bonnano and Constance, 2001). This problem arises because commodity chain research tends to focus upon the impact of industrialization on agro-food labour processes. It thus neglects the role of retailers, restaurants, nutritionists, market researchers, advertisers and other actors located closer to the consumption end of the chain (Dixon, 1999). As a consequence, there is little theoretical space to discern much deviation from the precepts of 'capitalist ordering' (either on the part of producers or consumers).

Building on this criticism, Whatmore and Thorne (1997, 288) suggest that the portrayal of commodity chains as 'systemic' and 'logical' tends to downplay miscellaneous forms of agency bound into material fabrics. In a discussion of 'fair trade' coffee, the authors propose that food networks must be conceptualized as composites of the various actors that go into their making. In doing so they argue that natural and technological entities as well as social actors be granted autonomy and causal powers. That is, multiple forms of agency should be considered when accounting for the establishment of (heterogeneous) relationships in food commodity chains (Whatmore and Thorne, 1997).

Whatmore and Thorne (1997) claim that a greater understanding of agro-food networks can be gained if food chain analysts engage with actor-network theory (ANT) – an approach that sees both human and non-human actors as holding the ability to consolidate and disrupt network relationships. They believe this theory has the potential to achieve a more thorough understanding of how networks and their socio-natural constituents hang together through time and space. Unlike political economy which tends to see organizational forms (such as chains or networks) as emblematic of already stabilized power relationships, actor-network theory 'problematises global reach, conceiving of it as a laboured, uncertain, and above all, contested process of "acting at a distance"' (Whatmore and Thorne, 1997, 290). ANT thus requires each network or chain to be studied in its own particularity and complexity (cf. Latour, 1996).

In this view, networks are complex because they arise from interactions between differing entity types (Latour, 1987). Entities coalesce, exchange properties, and (if the network is successfully consolidated) stabilize joint actions in line with overall network requirements (Latour, 1999). It is in this context that Callon (1991, 133) defines a network as 'a coordinated set of heterogeneous actors which interact more or less successfully to develop, produce, distribute and diffuse methods for generating goods and services'. An emphasis on heterogeneity implies that 'impurity is the rule' (Callon, 1991, 139). For Whatmore and Thorne (1997, 291–2),

> people in particular guises and contexts act as important go-betweens, mobile agents weaving connections between distant points in the network...But, insists [actor-network theory], there are a wealth of other agents, technological and 'natural', mobilised in the performance of social networks whose significance increases the longer and more intricate the network becomes...such as money, telephones, computers, or gene banks; objects which encode and stabilise particular socio-technological capacities and sustain patterns of connection that allow us to pass with continuity not only from the local to the global, but also from the human to the non-human.

That is, networks and commodity chains inevitably mobilize a multiplicity of social, natural and technological actors. The longer the networks and chains, the greater mobilization is likely to be.

Instead of the simplified world of capitalist ordering, we here encounter complex arrangements that comprise multiple rationalities, ordered in a variety of ways according to mixtures of entities assembled within the networks. However an emphasis on the heterogeneous quality of network relationships does not necessarily imply that each chain or network is unique (a uniqueness that is determined only by the combination of heterogeneous elements). Networks are rarely performed in radically new or innovative ways but rather incremental changes lead to 'new variations' on 'old themes'. Because network 'orders' tend to reflect widely dispersed 'modes of ordering' (Law, 1994), we see patterns and regularities in network relationships. Modes of ordering – which can be conceptualized as discursive frameworks holding together knowledge about past performances of network relations – are 'instantiated' and stabilized in given networks arrangements. Networks perform multiple 'modes of ordering', which influence the way actors are enrolled and how they come to be linked with others (Whatmore and Thorne, 1997, 294).

Because the notion of 'mode of ordering' is used in ANT theory to link discourses to networks it perhaps provides one means of establishing a connection between commodity chain and culture. However, ANT is less concerned with discourse than with non-human 'bits and pieces' that hold discursive formations together. It therefore tends to render ordering processes and network cultures in rather simplified terms. Having uncovered considerable socio-natural complexity in food commodity chains, for example, Whatmore and Thorne (1997) identify only two ordering modes: one that arranges materials according to a rationality of 'enterprise' and another that emphasizes the spatial 'connectivity' of entities and resources. Given that food networks come in many shapes and sizes, this two-fold typology seems unduly restrictive.[1]

A broader array of ordering modes can be found in conventions theory. Like ANT, this approach also proposes that network configurations stem from processes of negotiation between differing entities and discursive formations. However conventions theory pays more attention to the different modes of ordering that operate across contrasting cultural formations. It reconceptualizes modes of ordering as 'repertoires of justification' and proposes that repertoires have arisen historically to evaluate actions taken in the name of a 'common' or 'collective' good (Boltanski and Thevenot, 1991). Because it situates repertoires in varied cultural contexts, conventions theory is able to identify a broader range of modes of ordering than usually is evident in actor-network studies (cf. Lamont and Thevenot, 2000).

Conventions theory proposes that the heterogeneous arrangement of any particular network can be linked via repertoires of justification to the surrounding cultural or discursive context. In general terms, Thevenot *et al.* (2000) identify the following conventions as salient in providing this linkage:

- 'market performance', which evaluates worth based on the profitability, price or economic value of goods and services in a competitive market;
- 'industrial efficiency', which leads to evaluations based on long-term planning, growth, investment and infrastructure provision;
- 'civic equality', in which the collective welfare of all citizens is the evaluatory standard;
- 'domestic worth', in which value is justified by local embeddedness;
- 'inspiration', which refers to evaluations based on passion, emotion or creativity;
- 'reknown' or 'public knowledge', which refers to recognition, opinion and general social standing; and, lastly,
- 'green' or 'environmental' justifications, which consider the general good of the collective to be dependent upon the general good of the environment.[2]

Thevenot *et al.* (2000) argue that these justificatory repertoires exist in various combinations in all cultural contexts and serve to 'enable' and 'constrain' networking possibilities. In other words, they provide 'environments of action' in which network 'bits and pieces' are put together (Storper and Salais, 1997).

Whether or not the convention types listed above are exhaustive, they potentially extend the range of ordering processes thought to operate within food networks. Moreover, they allow us to specify the contrasting sets of linkages that might be established between particular conventions and particular arrangements of materials. As Lamont and Thevenot (2000, 7) argue, various evaluatory criteria will be employed to link network entities in particular ways: 'the treatment of persons (as customers) and things (as merchandise) that is required for market valuation is quite different from their treatment as experts and techniques that is required for an evaluation in terms of efficiency'.

The final shape of any network can therefore be seen to both reflect and enshrine convention-brokering processes. As differing conventions are negotiated, entities and relations will be integrated in line with the convention 'mix'.

In applying this perspective to the food sector we must consider how mixtures of conventions interact with combinations of natural, social and technological resources. Networks where modes of ordering reflect civic and domestic conventions will align a different set of materials and actors to those based on industrial criteria. Moreover, the brokering of conventions will extend from producers to consumers so that the shape and composition of food commodities will be decisively shaped by agreements enshrined within the material composition of commodity chains.[3]

Conventions theory thus finds a more complex interaction between cultural repertoires and the consolidation of heterogeneous materials within networks than either political economy or ANT. 'Modes of ordering' come in a variety of forms and reflect both the composition of the network and the cultural repertoires woven into its composition. In the remainder of the chapter, we consider the interaction between heterogeneous networks and conventions in the food sector by investigating two culinary networks in which different evaluatory criteria link the enrolled elements in distinct ways. We conclude with some thoughts on the globalization of cultural conventions within food networks.

Fast food culture

The story of McDonald's is well known and we will not repeat it at length here (see Love, 1986; Schlosser, 2001 for full, yet contrasting accounts). The company was born in the United States in the years following the Second World War, at a time of rising wages, an expanding birth rate, suburbanization, mass ownership of the motor car and increased leisure time. Eating out became a standard pastime as hamburgers and other fast foods met the new suburban requirement for 'convenience, efficiency and predictability in...food preparation and consumption' (Rifkin, 1992, 260). Jakle and Sculle (1999, 143) similarly argue that fast food sprang from a cultural milieu of 'competition, the quest for volume, the sense of urgency about service, and the symbolic role of the pampered consumer'. The hamburger in particular came to reflect prevailing cultural aspirations, as its 'capacity for speedy preparation with uniformly satisfactory results... mesh[ed] well with...demands of consumer and entrepreneur alike' (Jakle and Sculle, 1999, 144; Schlosser, 2001).

It was in this context that Maurice and Richard McDonald opened their first restaurant in Pasadena, California. The first McDonald's 'drive-in' sold mainly hot dogs to car-bound customers. After the success of this venture the brothers moved to San Bernardino where they opened a larger and even more successful 'drive-in'. However, the business was beset with problems, including high labour turnover: in a tight labour market the brothers experienced continuing difficulties in recruiting new workers. In the late 1940s, they closed the 'drive-in' and opened a new type of restaurant based on a refined system of food delivery. Under the 'Speedee Service System' the McDonald brothers dispensed with

> everything that had to be eaten with a knife, spoon, or fork.... [They] got rid of their dishes and glassware, replacing them with paper cups, paper bags, and paper plates. They divided the food preparation into separate tasks performed by different workers. To fill a typical order, one person grilled the hamburger; another 'dressed' and wrapped it; another prepared the milk shake; another made the fries; another worked the counter. For the first time, the guiding principles of a factory assembly line were applied to a commercial kitchen.
>
> (Schlosser, 2001, 20)

By employing an 'assembly-line' process the McDonald brothers were able to diminish labour requirements but could still deliver large quantities of burgers at low cost: 'a 1.6-ounce hamburger, 3.9 inches in diameter, on a 3.5 inch bun with .25 ounces of onion sold for 15 cents – a standardized product of high quality but also low price' (Jakle and Sculle, 1999, 141). In other words, the assemblage of heterogeneous elements within the McDonald's restaurant both reflected and refined a cultural convention of industrial efficiency.

The initial success of the 'Speedee Service System' attracted the attention of a travelling milkshake mixer salesperson named Ray Kroc. He apparently was impressed by the efficiency of the operation. Viewing the enterprise 'through the eyes of a salesman' (quoted in Schlosser, 2001, 67), Kroc anticipated substantial market growth. The McDonald's brothers, however, had no plans to extend the system beyond the one restaurant thus Kroc acquired the franchise and embarked upon an expansion of McDonald's restaurants across the US. Initially restaurants were located in suburban locations but as competition increased McDonald's moved into the cities. Following the company's flotation on the stock market in the 1970s much expansion has taken place overseas. By the mid-1990s McDonald's comprised 25,000 restaurants in 120 countries and global earnings stood at around $11 billion.

According to Ritzer (1996), Kroc achieved a 'revolution' in the fast-food sector by assembling a network dominated by economic efficiency. Importantly, the convention of efficiency is stabilized in a range of non-human technologies: 'the food in a McDonald's outlet is prepared by the use of timing mechanisms, beeping signals, pre-measured quantities, and computers submerged in the cooking oil that fry foods to uniform specifications' (Fantasia, 1995, 208). The same point is visible in Schlosser's (2001, 66) account:

> robotic drink machines selected the proper cups, filled them with ice, and then filled them with soda. Dispensers powered by compressed carbon dioxide shot out uniform spurts of ketchup and mustard. An elaborate unit emptied frozen french fries from a white plastic bin into wire mesh baskets for frying, lowered the baskets into hot oil, lifted them a few minutes later and gave them a brief shake, put them back into the oil until the fries were

perfectly cooked, and then dumped the fries underneath heat lamps, crisp and ready to be served. Television monitors in the kitchen instantly displayed the customer's order. And advanced computer software essentially ran the kitchen, assigning tasks to various workers for maximum efficiency, predicting future orders on the basis of ongoing customer flow.

Heterogeneity is orchestrated in line with an industrial convention so that the whole process of food delivery is engineered in line with a logic of efficiency. This logic also is evident in processes of food preparation. Because the food arrives at the restaurant

> pre-formed, pre-cut, pre-sliced and 'prepared' [there is] usually no need [for the workers] to form the burgers, cut the potatoes, slice the rolls, or prepare the apple pie. All they need to do is, where necessary, cook, or often merely heat the food and pass it on to the customer.
>
> (Ritzer, 1996, 103)

In the McDonald's system the industrial convention is materialized in the food preparation processes. This materialization extends to the restaurants themselves, which are designed to exact specifications, wherever they might be. McDonald's endeavours to make the consumption experience as repetitive as possible, with symbols, signs, colours, layouts all repeating a basic formula: ' "kitchen" visible in the background, tables and uncomfortable seats, prominent trash bins, drive through windows and so on' (Ritzer, 1996, 81).

While Ritzer's account usefully illustrates how an industrial convention is embedded in the heterogeneous materials that comprise the McDonald's food delivery system, his emphasis on an economic or industrial rationality neglects those network components that genuinely reflect the demands of consumers, and therefore the chain's 'market worth' (Kellner, 1999; Probyn, 1998, 2000; Smart, 1994, 1999). As Gottdiener (1997, 132) suggests:

> fast food outlets are successful because they offer an easy solution to the method of purchasing food that depends little on spoken language, on the interpretation of the menu or personal relations with the waitress/waiter, as happens in other restaurants. These and other themed environments, with their overendowed, instructive sign systems are fun places to be because they minimise the work we need to do for a successful interaction.

When entering McDonald's consumers need little prior knowledge of the consumption experience. Given McDonald's levels of marketing expenditure, consumers already will be familiar with the brand. Highlighting the importance of the 'reknown' convention in the network, Schlosser (2001, 5) notes that 'customers are drawn to familiar brands by an instinct to avoid the unknown. A brand offers a feeling of reassurance when its products are always and everywhere the same'. By repeating a basic formula, McDonald's transforms 'dining out' into a democratic process – anyone can (afford to) do it – and thereby expands its market.

It is arguable that we see here a civic convention in operation in which all consumers are equal before the food delivery system. Paradoxically it is the rigorous process of efficiency described by Ritzer (1996) that allows this civic convention to work: not only is the food relatively cheap, but also standardization allows consumers to adapt consumption spaces to their own requirements. Thus Kellner (1999) notes that in Taiwan restaurants have become study spaces for students, while in China and Russia they act as up-market eating establishments that provide welcome antidotes to the rather drab restaurants that surround them. McDonald's is therefore able to uphold a civic convention in varied spatial contexts. However, it is clear that the civic convention is closely tied to market worth and this appears to place limits on McDonalds' ability to mobilize a civic conception of the 'common good'. It is more likely to assert a 'common good' via market criteria.

In short, McDonald's skilfully aligns heterogeneous elements to deliver a standardized cuisine that meets a range of consumer aspirations. The arrangement encodes conventions of industrial efficiency and civic equality within the heterogeneous fabric of the food delivery system. The system can be exported into differing socio-economic and cultural circumstances so that consumers bring individual civic repertoires to the consumption experience. However the efficient nature of the system means that while these repertoires can be accommodated, they have little impact on the overall functioning of the food delivery process. That is, the civic convention is constrained by industrial and market criteria. As Fantasia (1995, 235) concludes in his study of fast food in France: 'with standardisation the hallmark of the fast food business, there is little room for the restaurant itself to change and develop over time in relation to the people who inhabit it'.

Slow food culture

According to Urry (2000, 43), 'global flows engender multiple forms of opposition to their various effects'. Emergent opposition to McDonald's takes a variety of forms including direct attacks on restaurants during the now frequent anti-globalization demonstrations. Our second case study represents a more long-standing form of opposition to this global network. It concerns Slow Food, a consumer movement established in Italy during the mid-1980s in direct response to the opening of one of the first McDonald's restaurants in Italy, in the famous Piazza di Spagna in Rome. The opening of this restaurant raised the possibility that traditional Italian eating habits might be under renewed threat from Americanized fast food. As part of the ensuing protest the food writer Carlo Petrini gathered chefs, authors, journalists and other intellectuals to discuss the most effective means of countering the spread of fast food in Italy. This first meeting in 1986 gave birth to a new consumer movement – Slow Food – which was to be devoted to the promotion of an 'anti-fast food' culture.

The initial aspiration of Slow Food was a celebration of cultural connections surrounding local cuisines and traditional products. The movement would target discerning consumers in order to heighten their awareness of 'forgotten' cuisines and the threats that these 'worlds of food' currently face. In this way it was hoped that new markets for traditional foods could be created. The main means of reaching potential consumers was to be a publishing house (Slow Food Editore) which would disseminate informed, interesting and accessible material on previously unknown or neglected foods. It was also intended that a network of local groups would be established in order to identify foods that are central to cuisines, to locate ingredients suppliers for these foods and to establish links to restaurants which would facilitate their consumption. Local groups would undertake activities explicitly aimed at tying actors more closely together in order to strengthen local cuisines. In other words, the network would give substance to the convention of domestic worth that is strongly present in Italian society.

The first edition of the movement's magazine, *Slow*, explicitly sought to oppose the spread of McDonald's and the other fast food chains. *Slow* (1995, page 1, Edition 1) claimed that the organization stood in opposition to the 'folly of fast life'. It proclaimed the need to nurture 'gentleness, pleasure, knowledge, care, tolerance, hedonism, balsamic calm, lasting enjoyment...culinary traditions ...' (ibid.). The symbol of the snail was adopted as the movement's logo. As Petrini (1986) explained:

> it seemed...that a creature so unaffected by the temptations of the modern world had something new to reveal, like a sort of amulet against exasperation, against the malpractice of those who are too impatient to feel and taste, too greedy to remember what they had just devoured.

As the adoption of this symbol suggests, the emphasis in Slow Food is upon the need to decelerate the food consumption experience so that alternative forms of taste can be (re)acquired. However, Slow Food also has spatial significance: the movement is concerned about a rupture between spaces of production and spaces of consumption and seeks to bring consumers closer to spatially embedded foods. It also wishes to reassert the natural bases of food production (seasonality, ecological content) and the role of cultural context (tacit knowledge, culinary skills). In short, it wishes to embed food in territory.

Slow Food's main concern is thus for 'typical' or 'traditional' foods. According to Torquati and Frascarelli (2000, 343), 'typicality' in the Italian context is determined by: 'historical memory (the product is associated with the history and with the traditions of the place of production), geographic localisation (influence of the pedoclimatic environment), [and the] quality of raw materials and techniques of preparation'. However, while it attempts to bolster these components of local cuisines, Slow Food also recognizes that local and regional food products are disappearing because they are *too* embedded in local food cultures and ecologies: they are not easily extracted and sold into modern food markets (for either cultural or ecological reasons, they often cannot travel the long distances covered by McDonald's burgers). So the movement attempts to attract consumers to traditional products by emphasizing aesthetic qualities. The magazine *Slow* promotes a highly aestheticized approach to consumption in lavishly illustrated articles. Slow Food Editore also publishes glossy consumer food guides that provide information on 'slow food' outlets. The most well-known, *Osterie d'Italia*, identifies typical restaurants in all the Italian regions thereby giving new consumers (e.g. tourists) the opportunity to engage with previously hidden, but long-standing, local foods. This process of aestheticization indicates the importance of the 'reknown' convention in the Slow Food culinary network.

While its roots are firmly within Italian food cultures, Slow Food seeks to promote 'typicality' much further afield. In 1989 it formally launched itself as an international movement and has subsequently spread to around forty countries (at the time of writing there are 70,000 members worldwide and offices in New York, Paris and Hong Kong). Local Slow Food groups are organized at the regional level into 'convivia'. Although the majority (over 300) are located in Italy, convivia now are operating in such differing national contexts as Australia, Brazil, India and the United States. Essentially, a convivium is a consumer club made up of people who wish to 'cultivate common cultural and gastronomic interests' (Slow Food, 1998). The definition of the local convivium area is given by cultural and culinary distinctiveness so that each group is charged with promoting a particular local cuisine. Convivia usually undertake the following activities: identifying restaurants that enshrine the principles of 'slowness' (mostly those offering a good selection of regional dishes and wines); organizing tastings of typical foods and talks by speciality producers and others on gastronomic issues; and promoting an appreciation of local foods in schools and other public institutions. In the process convivia highlight the creativity and tacit knowledge that reside in local cuisines. In this respect they promote a convention of inspiration.

In general, Slow Food seeks to build up cultural diversity by establishing close associations between local cuisines and local systems of production. In so doing, it conjures up collectives in which interactions between both people and nature are intense and close. In seeking to build these close associations the movement has come to recognize that many (ecologically sensitive) local producers and processors are precariously connected to consumers. It has thus decided to initiate more direct action in the production sector through a scheme called the *Ark of Taste* (Slow Food, 2000). Along with the usual activities oriented to the dissemination of knowledge about endangered products, the *Ark* project aims to set up another local group structure ('praesidia') which will be encouraged to identify producers in need of support, develop appropriate support measures (e.g. new marketing channels), and raise funds in order to put these measures in place. In this endeavour, Slow Food is beginning to work like an extension agency whose activities are tailored to producers and processors of typical products. In providing support to producers, Slow Food hopes to promote ecological diversity in local areas in line with the ecological convention.

The Slow Food network thus provides a useful comparator to McDonalds. Where McDonalds imposes a standardized format upon each locality, Slow Food encourages and supports multiplicity; where McDonald's is based upon the dissemination of a simple formula (e.g. 'The Speedee Service System'), Slow Food is built on an appreciation of diverse food cultures. However, there is a recognition in Slow Food that local diversity cannot simply be asserted as an alternative. Local cuisines and their constituent products first must be rendered transparent and made available to a wide number of potential consumers (this is done through the Slow Food publishing house). Second, cuisines must be protected by creating links between producers and consumers (this activity takes place through the 'convivia'). Finally, producers must be economically enabled to remain in existence (such efforts are made under the 'Ark of Taste').

In undertaking these activities Slow Food condenses a set of conventions found more widely in surrounding cultures of food (notably in Europe but also elsewhere): an appreciation of domestic worth (e.g. close spatial linkages between place of production and place of consumption); an attachment to food as a source of creativity and inspiration (e.g. in production, processing and preparation); civic equality (e.g. everyone should have access to a distinctive and local food culture and should be provided with the opportunity to develop new tastes); and environmental connectedness (e.g. the diversity of foods is a necessary accompaniment to biodiversity). While Slow Food refines and condenses these conventions from its cultural context, it adds in a distinctive characteristic: 'reknown'. It does this by promoting a highly aesthesticized culture of food in which the local, the traditional and the typical come to be seen as possessing considerable cultural value, even to consumers who understand little about the surrounding cuisine culture. In displaying food in this way, Slow Food not only seeks to provoke consumers into a reassessment of the traditional and the typical but also to generate a set of conventions which might obstruct the further 'McDonaldization' of food.

Conclusion

The networks described in the previous sections assemble heterogeneous elements in line with the contrasting mixtures of conventions that can be found within differing social formations. McDonald's condenses aspects of an 'Americanized' food culture to distribute a standardized and ubiquitous product via an industrialized system. While it may tailor products to local circumstances to a limited degree (e.g. selling pizzas in Italy, few meat products in Muslim countries or providing luxurious restaurant fittings in Monte Carlo), its strength is based on an economically efficient mode of food delivery which dispenses both cheap food to consumers and continuing profits to its shareholders. Slow Food, by contrast, is a consumer movement that promotes diversity in food production and consumption processes in order to safeguard the cuisine diversity that is found in Italy and other European societies. It therefore seeks to highlight connections between cuisines and regional natures and cultures and to strengthen markets for locally embedded products.

The two cases illustrate the different ways that global networks operate through space and time. McDonald's applies a uniform set of principles and seeks to turn all its network spaces into expressions of a single entity – 'McDonald's'. It also tends to efface the past in favour of a continuing present, a 'fast' present that can, again, simply be characterized as – 'McDonald's'. Alternatively, Slow

Food ties together a host of cuisines within sets of relations that give aesthetic expression to spatial diversity. It also shows concern for multiple times: it seeks to reinvigorate long-established regional and local cuisines and wishes to maintain these so that gastronomic connections are extended from the past to the future. Where McDonald's 'economizes', Slow Food 'culturalizes': McDonald's combines civic, market and industrial conventions in ways that ensure the dominance of economic criteria, while Slow Food combines civic, ecological and market conventions in ways that favour cultural and environmental criteria. The outcome in the first is a set of products which enshrine the principle of economic efficiency; the second gives rise to more culturally and environmentally embedded forms of production and consumption.

Such space–time effects indicate the obvious need to critically evaluate the networks one against the other. Clearly any evaluation could begin to consider their differential impacts on surrounding societies, cultures and ecologies. It might point out that McDonald's seems to 'externalize' many of the most significant interactions between food and environment. Because it links heterogeneous entities within instrumental modes of evaluation, 'the low price of the fast food hamburger does not reflect its real cost' (Schlosser, 2001, 261). Aspects of the production/consumption relationship (notably the health effects of the fast diet – see Vidal, 1997) are therefore displaced so that 'the profits of the fast food chains have been made possible by the losses imposed on the rest of society' (Schlosser, 2001, 261).

Slow Food, alternatively, appears to encourage some 'internalization' of costs within economic processes. In this network, market and industrial conventions are harnessed to those that highlight the cultural and environmental benefits of food. But this 'internalization' of cost means that many of the 'slowest' foods are relatively expensive. Thus, the Slow Food approach requires an 'aestheticization' of typical foods (i.e. an increase in their 'reknown') in order to attract those consumers who are willing (and able) to look beyond price to a much broader set of criteria. The apparent success of Slow Food in establishing an extensive global network based on an aestheticized food culture seems to indicate that a growing number of (predominantly) middle-class consumers are disposed to assessing food in this fashion. However, upmarket constructions of slow foods may restrict their market reach, thereby weakening the movement's ability to challenge the global expansion of fast food.

Whatever critical stance is taken, the success of each network in mobilizing complex relays of heterogeneous entities such that their length is constantly extended indicates that both are able to coexist in the food sector. This state of coexistence points to a possible fragmentation of production/consumption relations and indicates that 'flows of food' may be proceeding in different directions simultaneously. It also serves to substantiate Urry's (2000, 43) claim that the activities of global networks (both multinational and oppositional) currently serve to 'detotalise' national social forms thereby increasing the 'fluidity' of socio-economic life even further. In different ways McDonald's and Slow Food generate global flows that reach across national borders and cultures. Both cases highlight the existence of new (heterogeneous) interactions between networks and territories and suggest that spatial complexity will increase as global networks interact with discrete social formations to shape the practices of producers and consumers in diverse cultural contexts.

Notes

1 A similar criticism might be made of Law's (1994) study, which provides the basis for Whatmore and Thorne's (1997) two types.
2 In certain circumstances – such as deep ecology or animal rights – any distinction between collective and environment breaks down as the former incorporates the latter.
3 For further insights on the relationship between negotiation and commodity network composition, see Cook *et al.*, 2000; Murdoch and Miele, 1999; Murdoch *et al.*, 2000; Wilkinson, 1997.

References

Amin, A. and Thrift, N. (1995) Institutional issues for the European regions. *Economy and Society*, 24, pp. 121–43.

Boltanski, L. and Thevenot, L. (1991) *De la Justification: Les Economies de la Grandeur*. Paris: Gallimard.

Bonnano, A. and Constance, D. (2001) Corporate strategies in the global era: mega-hog farms in the Texas panhandle region. *International Journal of Sociology of Agriculture and Food*, 9, pp. 5–28.

Bonanno, A., Busch, L., Friedland, W., Gouveia, L. and Mingione, E. (eds) (1994) *From Columbus to ConAgra: The Globalization of Agriculture and Food*. Lawrence: University of Kansas Press.

Bridge, G. (1997) Mapping the terrain of time–space compression: power networks in everyday life. *Environment and Planning D: Society and Space*, 15, pp. 611–26.

Busch, L. and Juska, A. (1997) Beyond political economy: actor-networks and the globalization of agriculture. *Review of International Political Economy*, 4, pp. 688–708.

Buttel, F., Larson, O. and Gillespie, Jr. G. (1990) *The Sociology of Agriculture*. London: Greenwood Press.

Callon, M. (1991) Techno-economic networks and irreversibility. In Law, J. (ed.), *A Sociology of Monsters: Essays on Power, Technology and Domination*. London: Routledge.

Castells, M. (1996) *The Rise of the Network Society*. Oxford: Blackwell.

Cook, I. and Crang, P. (1996) The world on a plate: culinary culture, displacement and geographical knowledge. *Journal of Material Culture*, 1, pp. 131–54.

Cook, I., Crang, P. and Thorpe, M. (2000) Regions to be cheerful: culinary authenticity and its geographies. In Cook, I., Naylor, S., Ryan, J. and Grouch, D. (eds), *Cultural Turns/Geographical Turns: Perspectives on Cultural Geography*. London: Prentice Hall.

de Janvry, A. (1981) *The Agrarian Question and Reformism in Latin America*. Baltimore: John Hopkins University Press.

Dickens, P., Kelly, P., Olds, K. and Yeung, H. (2001) Chains and networks, territories and scales: towards a relational framework for analysing the global economy. *Global Networks*, 1, pp. 89–112.

Dixon, J. (1999) A cultural economy model for studying food systems. *Agriculture and Human Values*, 16, pp. 151–60.

Fantasia, R. (1995) Fast food in France. *Theory and Society*, 24, pp. 201–43.

Finklestein, J. (1999) Rich food: McDonald's and modern life. In Smart, B. (ed.), *Resisting McDonaldisation*. London: Sage.

Friedland, W. (1984) Commodity systems analysis: an approach to the sociology of agriculture. In Schwarzweller, H. (ed.), *Research in Rural Sociology and Development*. Greenwich, Connecticut: Jai Press.

Friedland, W. (2001) Reprise on commodity system methodology. *International Journal of Sociology of Agriculture and Food*, 9, pp. 82–103.

Friedland, W., Barton, A. and Thomas, R. (1981) *Manufacturing Green Gold*. New York: Cambridge University Press.

Goodman, D. (1999) Agro-food studies in the 'Age of Ecology': nature, corporeality, bio-politics. *Sociologia Ruralis*, 39, pp. 17–38.

Goodman, D. and Redclift, M. (1991) *Refashioning Nature: Food, Ecology and Culture*. London: Routledge.

Goodman, D. and Watts, M. (eds) (1997) *Globalising Food: Agrarian Questions and Global Restructuring*. London: Routledge.

Goodman, D., Sorj, B. and Wilkinson, J. (1987) *From Farming to Biotechnology*. London: Routledge.

Gottidiener, M. (1997) *The Theming of America*. Boulder, CO: Westview.

Hannerz, U. (1992) *Cultural Complexity*. New York: Columbia University Press.

Ilbery, B. and Kneafsey, M. (1998) Product and place: promoting quality products and services in the lagging regions of the European Union. *European Urban and Regional Development Studies*, 5, pp. 329–41.

Jakle, J. and Sculle, K. (1999) *Fast Food: Roadside Restaurants in the Automobile Age*. London: Johns Hopkins University Press.

Kellner, D. (1999) Theorizing/resisting Mcdonaldization: a multiperspectivist approach. In Smart, B. (ed.), *Resisting McDonaldization*. London: Sage.

Kirsch, S. (1995) The incredible shrinking world: technology and the production of space. *Environment and Planning D: Society and Space*, 13, pp. 529–55.

Lamont, M. and Thevenot, L. (2000) Introduction: toward a renewed comparative cultural sociology. In Lamont, M. and Thevenot, L. (eds), *Rethinking Comparative Cultural Sociology: Repertoires of Evaluation in France and the United States*. London: Cambridge University Press.

Latour, B. (1987) *Science in Action*. Milton Keynes: Open University Press.

Latour, B. (1996) *Aramis, or the Love of Technology*. Cambridge, MA: Harvard University Press.

Latour, B. (1999) *Pandora's Hope*. Cambridge, MA: Harvard University Press.

Law, J. (1994) *Organising Modernity*. Oxford: Blackwell.

Love, J. (1986) *McDonald's: Behind the Arches*. New York: Bantam.

Lowe, M. and Wrigley, N. (1996) Towards the new retail geography. In Wrigley, N. and Lowe, M. (eds), *Retailing, Consumption and Capital: Towards the New Retail Geography*. London: Longman.

Lupton, D. (1996) *Food, the Body and the Self*. London: Sage.

Miele, M. (2001) Changing passions for food in Europe. In Buller, H. and Hoggart, K. (eds), *Agricultural Transformation, Food and Environment*. Aldershot: Ashgate.

Miller, D., Jackson, P., Thrift, N., Holbrook, B. and Rowlands, M. (1998) *Shopping, Place and Identity*. London: Routledge.

Montanari, M. (1994) *The Culture of Food*. Oxford: Blackwell.

Murdoch, J. (1995) Actor-networks and the evolution of economic forms: combining description and explanation in theories of regulation, flexible specialisation and networks. *Environment and Planning A*, 27, pp. 731–58.

Murdoch, J. (1998) The spaces of actor-network theory. *Geoforum*, 29, pp. 357–74.

Murdoch, J. (2000) Networks – a new paradigm of rural development? *Journal of Rural Studies*, 16, pp. 407–19.

Murdoch, J. and Miele, M. (1999) 'Back to nature': changing worlds of production in the food sector. *Sociologia Ruralis*, 39, pp. 465–83.

Murdoch, J., Marsden, T. and Banks, J. (2000) Quality, nature and embeddedness: some theoretical considerations in the context of the food sector. *Economic Geography*, 76, pp. 107–25.

Petrini, C. (1986) The Slow Food Manifesto. *Slow*, 1, pp. 23–4.

Probyn, E. (1998) Mc-Identities: food and the familial citizen. *Theory, Culture and Society*, 15, pp. 155–73.

Probyn, E. (2000) *Carnal Appetites: Food/Sex/Identities*. London: Routledge.

Reiter, E. (1991) *Making Fast Food: From the Frying Pan into the Fryer*. Montreal: McGill-Queens University Press.

Rifkin, J. (1992) *Beyond Beef: The Rise and Fall of the Cattle Culture*. New York: Dutton.

Ritzer, G. (1996) *The McDonaldisation of Society*. London: Sage.

Schlosser, E. (2001) *Fast Food Nation: The Dark Side of the All-American Meal*. New York: Houghton Mifflin.

Slow Food (1998) The convivia. *Slow*, 15, pp. 10–11.

Slow Food (2000) *The Ark of Taste and the Praesidia*. Bra: Slow Food Editore.

Smart, B. (1994) Digesting the modern diet: gastro-porn, fast food and panic eating. In Tester, K. (ed.), *The Flaneur*. London: Routledge.

Storper, M. and Salais, R. (1997) *Worlds of Production*. Cambridge, MA: Harvard University Press.

Thevenot, L., Moody, M. and Lafaye, C. (2000) Forms of valuing nature: arguments and modes of justification in French and American environmental disputes. In Lamont, M. and Thevenot, L. (eds), *Rethinking Comparative Cultural Sociology: Repertoires of Evaluation in France and the United States*. Cambridge: Cambridge University Press.

Torquati, B. and Frascarelli, A. (2000) Relationship between territory, enterprises, employment and professional skill in the typical products sector. In Sylvander, B., Barjolle, D. and Arfini, F. (eds), *The Socio-Economics of Origin-labelled Products in Agri-food Supply Chains*. Le Mans: INRA.

Urry, J. (2000) *Sociology Beyond Society: Mobilities for the Twenty-first Century*. London: Routledge.

Vidal, J. (1997) *McLibel: Burger Culture on Trial*. London: Macmillan.

Warde, A. and Martens, L. (2000) *Eating Out: Social Differentiation, Consumption and Pleasure*. Cambridge: Cambridge University Press.

Whatmore, S. (1994) Global agro-food complexes and the refashioning of rural Europe. In Amin, A. and Thrift, N. (eds), *Globalization, Institutions and Regional Development*. Oxford: Oxford University Press.

Whatmore, S. and Thorne, L. (1997) Nourishing networks: alternative geographies of food. In Goodman, D. and Watts, M. (eds), *Globalising Food: Agrarian Questions and Global Restructuring*. London: Routledge.

Wilkinson, J. (1997) A new paradigm for economic analysis? *Economy and Society*, 26, pp. 305–39.

6 Initiating the commodity chain

South Asian women and fashion in the diaspora

Parvati Raghuram

Introduction

Émigrés who have moved from the south to the north increasingly have been recognized as key actors in business networks: as crucial agents in global capitalism (Mitchell, 1995; Olds and Yeung, 1999; Yeung, 2000) who exert control at nodes in global commodity chains. However, most current discussions offer a reduced understanding of the role of women in such chains, ascribing to women little individual agency. Within the fashion commodity chain, South Asian women have played an important role in initiating networks of production. In highlighting South Asian women as key agents, this chapter challenges the limited ways in which their participation in such networks has so far been envisaged. Methodologically, I argue that focusing on personal as well as product biographies can enhance our understanding of the existential setting within which commodity chains operate and can help us to better understand the ways in which consumption and production are intricately linked.

Joining the ends of the global commodity chain

A critical analysis of the globalization of design, production, advertising, retailing, marketing and consumption of goods and services has been undertaken under the rubric of the global commodity chain (GCC) literature (Gereffi, 1994). The notion of a commodity chain traces the entire trajectory of the product, usually within a political economy of development perspective derived from world systems theory (Gereffi, 1994; Raikes *et al.*, 2000). A distinction is commonly made between producer-driven chains such as automobiles and aircraft and buyer-driven chains such as garments, fresh fruit and vegetables. However, as Leslie and Reimer (1999) have argued, research on GCCs has been dominated by the impetus of production. Even accounts of buyer-driven chains tend to adopt the moment of production as the origin, the essence of economic meaning and the moment of initiation of the commodity chain. Such a focus betrays the Marxist underpinnings of the GCC model.

Production tales dominate the GCC literature (Gereffi, 1999; Mather, 1999) particularly those initiated by transnational corporations. Gereffi's (1999, 43) analysis of the apparel commodity chain, for example, outlines the ways in which the main leverage is exercised by 'retailers, marketers and manufacturers through their ability to shape mass consumption via strong brand names and their reliance on global sourcing strategies to meet this demand'. Case studies within the GCC literature have included primary products such as fish, fruits and flowers (Gibbon, 1997; Gwynne, 1999; Hughes, 2000) or clothing (Crewe and Davenport, 1992; Dicken and Hassler, 2000; Van-Grunsven and Smakman, 2001). In both sets of industries significant continuations and reconfigurations of colonial production regimes are visible, however narratives of marketing and consumption commonly mask these production processes. Because fish, fruit and flower products are usually minimally processed, places of production often appear in marketing strategies via images which seek to construct idealized 'traditional' societies (see Cook *et al.*, Chapter 9, this volume; Coulson, Chapter 7, this volume). However the clothing commodity chain differs to the extent that marketing campaigns usually must erase images of processing in order that they can be replaced with images and narratives of consumption.[1]

For my purposes in this chapter, four aspects of the mode of analysis adopted by the GCC literature are worth noting. First, although GCC analyses are critically concerned with issues of power and the mechanisms through which this power is played out, control is always seen to rest with large producers. 'Mass consumption' is seen to be shaped by the interests of retailers while consumers display little evidence of individual agency. Second, the global scale of the analysis has only ultimately produced accounts of the activities of large global corporations. Small producers are generally absent from the story. For example, in Gereffi's (1994) analysis of the apparel industry the use of terminology such as 'driven' is suggestive of an arrangement under which major participants are able to initiate and energize very large systems of production and distribution. The analysis becomes limited to actors who are capable of driving such systems rather than those who participate in them. The degree of control inherent in the 'driving process' means that it is often global corporations that come to be studied.

Third, GCC approaches methodologically separate producers from consumers, casting producers only in that role and removing from view producers' participation in the consumption process. Questions about what agents who mobilize the production of clothing in faraway places *themselves* wear find no answers within this paradigmatic framework. The model is not adept at identifying factors influencing the design of products and rarely considers the ways in which 'consumer choice' might influence designs. Consumers often redesign products that are being sold by large organizations by reconfiguring the ways in which products are used (Du Gay *et al.*, 1997). In order to bring consumers back into focus, therefore, analyses need to take a step 'back' into design issues.

Finally, the GCC mode of analysis tends to view the 'Third World' only ever as a site of production. In attempting to highlight the ways in which global inequalities are produced and reproduced, 'Third World' states are positioned as producers for the so-called global market. In fact GCC accounts grossly oversimplify the nature of global markets, for the consumption practices of Third

World producers never appear. While a focus on power and inequalities potentially may be radical, removing agency via the erasure of Southern consumption serves to objectify the 'Third World' as a geopolitical entity. Residents of the Third World only become subjects through their role in the 'new' international division of labour.

Attempts to redress the productionist bias of the GCC literature have been primarily developed through culturalist interrogations of contemporary commodity culture. In geography, these accounts have been best developed by cultural geographers working on consumption sites, on commodities and on the processes of consumption (Cook *et al.*, 2000; Gregson *et al.*, 2000; Jackson, 1999). Most often, such writing has drawn upon research in cultural studies which argues that the meanings of products are produced/reconstituted at the point of consumption. Consumption thus becomes a key practice in the cultural economy circuit (Appadurai, 1986; Du Gay *et al.*, 1997; Johnson, 1996; Lash, 1997). This focus on consumption attempts to overturn the economic determinism of production studies models by arguing that culture is not just an effect but is also constitutive of social and economic formations. Cultural forms influence the ways in which the economy is represented: processes of representation and the meaning of representation (signs) exercise their own force and their own determinacy on our understandings of these formations. One of the strengths of this position is that it is more likely to recognize the spatio-temporal contingency of the products and processes we study. It aims to study meanings but rarely posits those meanings as universal.

Much of the work produced in this vein however privileges *product* biographies. For instance, Lash's (1997) research on the culture industries focuses on ten media products. His method involves an analysis of the 'transformations of cultural products as they move from one stage of production to the next: from production to distribution to sales to consumption, from one sector to another' (Lash, 1997, 4). Moreover, Lash traces these ' "backwards" in time, from consumption through marketing back through cultural intermediaries all the way to the various stages of design and original production' (ibid.). A focus on the products of consumer culture arises from the concern to reveal and explore contradictions between the exploitative production processes of capitalism and products' seamless appearance and symbolic value. Attention is paid to the processes whereby cultural artefacts conceal their origins ideologically (Johnson *et al.*, 2004). However although a focus on products provides a corrective to anthropological stances which had been primarily absorbed with social life rather than the social life of things, there are certain limitations to the object-centric direction which this research has taken.

One attempt to overcome an entirely object-centric focus and to foreground the role of the subject is provided by Latour's (1993) actor network theory (ANT). Not only are subjects seen to be incorporated through their role as actants in networks but also the notion of the subject itself is rewritten so that both humans and non-humans acquire agency. ANT has been utilized by cultural theorists to highlight the multiple forms of agency available to and mobilized by individuals acting at multiple nodes in order 'to develop, produce, distribute and diffuse methods for generating goods and services' (Callon, 1991, 139). For example, Lash (1997, 5–6) suggests that ANT offers a method of understanding 'how key actors recruit, not just other actors but mobilise the cultural products themselves and the affordances which they, and technologies of distribution, offer as forces in such struggles.'[2] However in focusing primarily on the agency of human and non-human entities, ANT is in danger of reducing complex production and consumption systems into a mechanistic framework. Concentrating on agents operating at nodes within the network and focusing solely on the links between them can disembed these systems from social frameworks and processes within which these links are set (Dicken *et al.*, 2001). The discursive framing of social relations and the pre-existing rationalities that underpin networks may both be sidelined in ANT analyses.

For me, a central difficulty with all the narratives reviewed here – GCC, cultural studies accounts and ANT – derives from their location within a particular form of commodity culture in which, Marx would argue, labour is ultimately alienated (Du Gay *et al.*, 1997; Lash, 1997). Whether examining products intended for an elite market or for mass consumption, existing accounts often distance the producer and consumer from each other. Those who produce are never seen to be consuming the goods they produce. A woman who may sew clothes at home, for example, is only ever positioned as a sweatshop worker producing for global markets. Forms of production and consumption which have been incompletely subsumed into the capitalist economy, such that the activities of women who produce their own clothes to wear rarely enter these debates. The forms of production which have come to be studied are themselves quite specific to urban, industrial and capitalist cultures. Men and women whose experiences cannot be formulated within these models have largely been ignored. Moreover, production for consumption cannot always be cast as a pre-industrial activity. Explicit recognition of the creativity and pleasure achieved through production challenges existing literatures which largely identify pleasure at the point of consumption. Focusing on those who produce in order to consume what they produce also provides an interesting methodological twist to existing stories. It entails the inclusion of personal biographies as well as product biographies and requires the study of circuits of production to also include the histories and geographies of producers. Critically, this mode of analysis might help to strengthen the political edge to studies of consumption and production (Hartwick, 2000). It enables the terms of women's engagement to be understood as 'simultaneously historically specific and dynamic, not frozen in time in the form of a spectacle' (Mohanty, 1991, 6).

Gendering the global commodity chain

The problems of separating production and consumption in the GCC become even more apparent when explicitly addressing questions of gender. Leslie and Reimer (1999, 409) have argued that a focus on the verticality of the chain obscures horizontal relations such as those of gender which 'underpin commodity chain dynamics' and mediate relationships across the chain. Where gender is addressed, all men are identified as primary economic actors. Narratives of production are masculine tales within which any identification of *individual* agents focuses upon men. The economic is thus over-determined as masculine and patriarchal.

A contrasting story is told within literature on gender and entrepreneurship. Marlow and Strange's (1994, 1) consideration of 'white' women entrepreneurs argues that entrepreneurship research has been 'largely gender blind'. 'Entrepreneurial theories...[are] created by men, for men, and...applied to men' (Holmquist and Sundin, 1989, 1). Women rarely appear as actors in this literature. If featured at all, they are positioned as unpaid family labourers – cultural and social reproducers who facilitate the 'real work' of economic production undertaken by men. Even gender-sensitive research on entrepreneurship (e.g. Owen, 1995) has tended to ignore the presence of self-employed women among ethnic minority groups in Britain, despite the fact that the proportion of ethnic minority women in self-employment is substantially higher than that of the white population (see Phizacklea and Ram, 1996).

Third World women[3] primarily appear in production tales as workers in the (not so) new international division of labour. Feminist interventions into the development literature have meant that the importance of women's labour in production sites in the Third World is now well established. A large number of studies have outlined women's role in export processing zones working in factories owned directly or indirectly by transnational corporations (Nash and Fernandez-Kelly, 1983) and as atomized homeworkers producing for the world market (Mies, 1982). Yet production and consumption are rarely linked. Du Gay *et al.*'s (1997) story of the Sony Walkman is illuminating in this regard. Photographs of rows of women working on an assembly line are juxtaposed with images of young men and women using the product at the consumption 'end' of the chain. Viewed together, the images display women's centrality to production process and to Sony's economic success, at the same time that their labour is made invisible in advertising and selling the product.

A (broadly) Marxist political economy perspective has come to dominate much of the gender and development (GAD) as well as the GCC literature. As a result, relations of exploitation tend to derive from the world system, and 'hardworking women' are primarily located in the Third World. An exception to this narrative occurs when the Third World women come 'home' – when they migrate to the metropoles and reconstitute themselves as the 'underclass' within the First World. What is common to the process of production is the persistent double oppression of race and gender and it is at the cross-section of these hierarchies that women workers of the world are placed. As Mohanty (1997, 20) has argued, 'ideologies of domesticity, femininity, and race form the basis of the construction of the notion of "women's work" for Third World women in the contemporary economy'.

Although the GAD literature provided an important corrective to earlier accounts, this mode of analysis too has for too long carried with it limited notions of women's economic agency. Few studies highlight the contradictions of women's positions within production sites, the social agency they have as individuals, or their participation in collectives such as trade unions (although see Ong, 1987). Neither do women themselves ever become consumers. The role of 'workers' is also the only place available to Third World women within the economic narrative.

While a number of studies have highlighted the importance of women as consumers (Nava, 1991), providing women with agency in the commodity chain, these women are usually conceived of as 'white'. A complex set of interactions between race, class and global political inequalities has meant that white women in the First World may often be the women with the most purchasing power. However, such women then only ever appear as consumers in narratives of the global economy, a representation that is ultimately as reductionist as representing all Third World women only as producers. In a deft double movement, contemporary economic narratives have also separated producers from consumers. Women are never both.

As I have argued above, envisaging producers as consumers can not only help to tie up the ends of the commodity chain but it can also provide one mode of entry for recognizing the multiplicity of positions that women may adopt within the GCC. Typically, producing for oneself – or self-provisioning – has been seen as an activity not of choice but of survival and as a residual category within the modern narratives of the global political economy. Those who are 'forced' to produce in order to consume are imagined both as 'traditional' and also often carry the burden of particular class positions. The picture of a woman sewing her own clothes commonly appears in the development literature where it is argued that self-provisioning is increasingly being adopted as a strategy by those who have been adversely affected by fall in income and increasing commodity prices due to Structural Adjustment Programmes. However the story of self-provisioning is much more complicated than representations of poor 'Third World women' might suggest. The following section turns to find a place for South Asian women at the nexus of production and consumption tales.

South Asian women and the consumption of clothing

South Asian women in the United Kingdom are primarily represented as reproducers (of culture, of tradition?), with limited participation in the economy. Their 'productive' role is recognized as an extension of roles performed by women in the 'Third World' as cogs in the capitalist wheel. Concomitantly, South Asian women are seen to have little purchasing power and their consumption patterns are of little interest to academics. By casting South Asian women primarily as reproducers, consumption is assumed to occur largely within the context of their reproductive role, undertaken in order to support the family, to feed the husband and to buy clothes for the children. Here, consumption is not for pleasure but is part of a duty, a part of women's caring roles. In this discourse the consumption of personal products may only be justified through the pleasure it provides others.

Significantly the choices that women make about what to consume, and what not to consume, also carry with them symbolic value. Theoretical frameworks that cast South Asian women as 'traditional' in the context of a binary division between 'tradition' and 'modernity' will also by extension see women as purchasing goods which are 'traditional', as well as based primarily on their racialized identity. In such analyses material cultures are seen to be 'preserved' through the consumption practices of diasporic women.

In contrast to these implied yet often unstated assumptions about South Asian consumption practices, the narratives produced by young women in the diaspora reveal quite different accounts. South Asian women foreground the significance of purchasing not for the family but also for oneself, and consuming not as part of caring for others but for pleasure. Mani's (2002) discussion, for example, disrupts the tradition/modernity binary by exploring the ways in which women in the United States mix and match Western and Asian forms of clothing, and problematizes paradigmatic constructions of the referent geopolitical entities. Such research successfully challenges stereotypical representations both of the clothes themselves and of those who wear them.

Importantly, this reconfiguration of consumption practices is not just occurring in diasporic spaces. Rather, as Munshi (2001) argues, middle class urban Indians also are negotiating modernity/tradition dichotomies. In particular, producers of visual media have become conscious that enlarging the range of feminine subjectivities will alter and increase consumption (ibid.). The 'modern' woman targeted by advertisers 'now in most cases combines both family and career, familial relationships and independence, selflessness and a little (long overdue) selfishness' (Munshi, 2001, 81). Advertisements emphasize the promise of individual pleasure and fuel an increasingly intensive investment in women's appearance. Products consumed by South Asian women draw upon both international styles and on 'international expertise' – and international origins or references of products are used to validate the usefulness and 'authenticity' of goods. While some advertisers manipulate tales of modernity to sell products, others retail images of 'Indianness' and emphasize the importance of that image within the context of new reconfigurations of gender in South Asia. As Munshi's (2001, 91) account suggests, 'the "marvellous me" persona wears the "international" look on the outside but is a real "home" girl at heart'. Emphasizing women's agency in advertisements thus not only increases the sales of 'modern' products but also serves to market goods with longer histories of consumption in India.

Reworkings of the modern/traditional binary may also be understood by considering shifts in fashion itself. An important example is provided by the salwaar kameez suit. While this dress form marks a British wearer as 'traditional', in India the salwaar kameez represents 'a transitional item of clothing spanning non-western and western fashion systems. It offers Indian women a choice of clothes to suit the practical requirements of different lifestyles and occasions, reflecting 'the desire of some Indian women to be westernised and cosmopolitan' (Craik, 1994, 35). Even if it is not a mark of westernization, it may well denote cosmopolitanism, signalling a shift from regional forms of clothing.

The persistence of the salwaar kameez as a dress form must then be understood not only in the context of the social changes in India, but also the changes faced by South Asians who have emigrated. South Asian women who participated in early waves of migration to the United Kingdom often continued to wear salwaar kameez. These forms of clothing were worn not

> as an explicit and overt political statement against racism, but because of a different sense of aesthetics, one that manages to exist in spite of the racialising lens … In other words, the racism and orientalising of South Asians [did not] kill or contain an alterity of aesthetics
>
> (Puwar, 2002, 77–78)

However women who wanted to wear salwaar kameez or saris found few stores in the United Kingdom from which to buy these clothes (see Puwar, in press). Thus, for many women, sewing one's own clothes provided the main route to wearing clothes of one's choice, a common enough practice not just among migrants but throughout the world.[4] The emergence of South Asian women's clothing shops was distinctive in that retailers served a small niche market, that differentiation of products to suit different aesthetics took longer and has depended on the growth of a market large enough to support these different tastes. Self-provisioning continued to be an important strategy for South Asian women – perhaps longer than for women wearing 'Western clothes' – because a commodified 'ready-mades' market in 'Asian' clothes emerged later. Most recently, the emergence of new niche markets to meet the demands of a population marked by a confident reconfiguring of ethnicity and also by class has lead to the mushrooming of differentiated products and the growth of new retailing outlets (Bhachu, 1995, 2000; Raghuram and Hardill, 1998).

The remainder of the chapter explores the ways in which one South Asian woman entrepreneur initiated a successful enterprise in order to meet her own clothing requirements. Having identified a growing demand for particular forms of clothing, she responded to this by initiating a production chain. Although she did not produce her own clothes, the impetus to produce arose from a recognition of her own consumption needs. This pattern undoubtedly is common in entrepreneurship – many ideas for business start up derive from personal experience. However few studies have recognized the importance of the process by which an unsatisfied customer becomes a producer who makes 'what they want', and subsequently comes to consume their own products.[5] Recognizing this pattern crucially highlights subjective elements in the production process and extends current research in which identity formation often is encountered only at the point of consumption.

The personal biography method is strategic on two accounts. First, it enables an understanding of the agency of South Asian women beyond simply participation in production cycles while at the same time making more complex the histories of women's consumption within the South Asian diaspora. Although the chapter privileges the existential setting of one woman's narrative, the discussion is relevant to the histories and geographies of other producers who share similar spaces and times. Second, the examination of a particular biographical narrative enables me to highlight multiple layers present in the economic relationships in which individuals engage and to thus draw together issues of power relations with those of agency.

Placing the story

The chapter now turns to the story of Malini, a 31-year-old East African Asian (Gujarati) businesswoman living in the Midlands.[6] Her experiences form part of a broader migratory pattern, extending back and through the histories of three continents. Asian immigration to East Africa preceded British colonial policies but was intensified and transformed through colonialism. Until 1922 Asians – and particularly Gujaratis – were taken to East Africa as indentured plantation labourers. After the abolition of the indentured system in 1922, subsequent migrations were largely voluntary and migrants filled the middle strata of colonial society in Kenya, Uganda and Tanzania, between the more prosperous and politically powerful white minority and the poorer rural African populations. This group of Asians, many of whom originated from rural areas, became engaged in business and administration in urban areas and acquired many markers of 'modernity': knowledge of English, urban behaviour and administrative skills.

While the Asian community in East Africa was never monolithic, capitalism operated to solidify their experiences in the context of greater differences which existed between them and other population groups (Africans and Europeans) in colonial East Africa (Bujra, 1992). Asians became positioned as a 'middle strata' of business people who conduited profits extracted from the African peasantry to the European colonial powers. Malini's family were late immigrants from India, having left Gujarat in 1948 and settled in Zambia, where they retailed children's clothing. Malini's father moved to Great Britain in 1954 before the large migration flows generated by political changes following the independence of many East African countries in the 1960s. He was joined by his wife and two children in 1958. Malini was born after the family settled in Leicester.

Malini's family story provides an insight into the historico-geographical complexities which lie behind the formation of diasporas. Diasporic populations are constituted not merely through the dispersion from the homeland but also through further dispersions and population movements. Furthermore these population movements are not necessarily conditioned by the return to the country from which they originally emigrated. 'Multi-locale' dispersions have been particularly central in the case of the South Asian diaspora. Diasporic populations thus retain linkages not only with 'home' but also with other parts of the diaspora.

The experiences and skills of East African Asians in Britain differ from some of those who have directly migrated from the Indian subcontinent. While a move to Britain was the first migratory experience for the latter group, East African Asians have considerable entrepreneurial experience and an urban base, as I have indicated. As a result, the British Asian population currently is very diverse and is marked by linguistic, religious, regional and sectarian differences. Furthermore, social polarization between and within Asian communities increased during the 1980s and 1990s, with indirect migrants emerging as real 'winners' (Metcalf *et al.*, 1996).[7] Higher economic class positions are also cross-cut by gender: typically, Asian men appear to have benefited more than Asian women.

The ultimate settlement of Malini's family in Leicester forms part of the story of broader migrant flows into the city. Indians and Pakistanis who moved to meet post-war labour shortages in the 1950s and early 1960s were followed by family reunion migrants. In the late 1960s and 1970s, the more numerous migrants from East Africa deepened the concentration of Asians, particularly in the Highfields and Belgrave areas (Nash and Reeder, 1993, 27). Although more recent patterns of suburbanization have led to a residential 'dispersal' (Byrne, 1998; O'Connor, 1995) retailing continues to be concentrated in central city wards, particularly along the Belgrave and Melton Roads. This area now represents the most important retail centre for Asians in the East Midlands and for Gujaratis nationally, offering a central focus for celebrations like Diwali.

Clothing and the diaspora

This section begins by sketching out the larger context of recent clothing trends, before returning to the details of Malini's participation in the fashion commodity chain. As is the custom and practice in the Indian subcontinent, British Asian men are, on the whole, less likely to wear 'Asian clothes' every day than Asian women are, although the use of turbans by Sikh men is an interesting exception to this pattern. Generalized patterns vary with religious affiliation and region of origin in the Indian subcontinent, as well as area of residence in the United Kingdom. For instance, Muslim men are more likely to a wear salwaar kameez than non-Muslim men. However, these patterns are continuously changing so that non-Muslim men too may now replace the Western suit with 'Asian' clothes for special occasions such as their own weddings. A shift to Asian clothing cannot be read as a direct response to the increased availability of such clothing but rather may be related to the ways in which certain garments are acquiring hegemonic status both in the diaspora and in the subcontinent. Thus the 'dhoti' may be replaced by a sherwani kurta and trousers at wedding receptions in South India in the same way that Western suits are being replaced by these garments in North India and in the South Asian diaspora. Media representations of clothing, and in particular the growing influence of fashion presented in Bollywood films have had a significant impact on all these patterns.

Asian women often wear 'Asian clothing', particularly for special occasions – for festivals or for social gatherings surrounding life course events, such as birth, marriage or death. The importance of changing dresses for different occasions interestingly highlights the performative aspect of culture – of the importance of dressing up, of knowing appropriate social codes and of recognizing the historically and spatially dynamic nature of these codes. Men appear to be more likely to wear a salwaar kameez in areas where there is a relatively substantial visible Asian presence, or when visiting such areas. For both men and women, 'keeping up' with fashion thus requires temporal as well as spatial sensibilities.

These codes also intersect with economic processes and with particular forms of class stratification. An increasing social stratification within British society (Hills, 1998) is also reflected (and perhaps even magnified) within the Asian community. An increasingly important form of differentiation has emerged alongside a growth in the Asian managerial and professional service-class, which has stimulated specific forms of consumption.[8] The clothes of professional women in particular often have a symbolic value greater than the material value of the product. This symbolic value arises out of meaning-making processes formed within reference systems bounded by class as well as by racialized and gendered differences. Forms of consumption become the tools not only for reflecting status but also the means for gaining and negotiating status across these social hierarchies (Bourdieu, 1984).

Doing fashion and having detailed knowledge of particular codes is most significant in women's fashion. For the diasporic population this involves a complex interaction with the fashion industry in the home country. Although clothing styles vary markedly across the Indian subcontinent, women's garments such as the sari and more particularly the salwaar kameez increasingly have gained hegemonic status both in the subcontinent and amongst diasporic populations (Tarlo, 1996). This trend has stimulated growth in markets for high quality salwaar kameez in both India and in the Indian diaspora. As a result the fashion garment industry has concentrated on these two forms of clothing or variants of them.[9]

An astute awareness of the trends I have described above led Malini to establish a business in 1985. As a member of the cohort to whom she sought to sell, Malini was able to recognize the complexities of differentiation and stratification, to identify the importance of consuming the 'right thing' and to be aware of the rapid increase in the numbers who could engage in particular forms of consumption. Malini's business was started at least in part because she was an unsatisfied customer. Further, however, she deliberately targeted a group of 'conspicuous consumers' who cut across cultural, racial and ethnic communities.

Malini identified a new, innovative market niche of 'Asian' fashion clothing – ready-to-wear salwaar kameez – which were products that she herself wanted to wear for social occasions, but which were not available in the mid-1980s. Initially, Malini had a small number of garments made to order which she then sold through a friend's retail outlet. The success of this venture encouraged her to enter into partnership with the owner of an existing business:

> I have always wanted to do clothes and I knew there was a potential for this industry ... at the time we started it was so new. People didn't talk about ready-mades then ... I knew it was going to take off in a big way so I got in at the right time.

Malini began retailing ready-to-wear salwaar kameez in order to capture the market that she identified as a consumer and in particular targeted Asians who appeared to have a higher purchasing power: 'the classy Punjabis and the Malawi Muslims…they have got investments here and there…are major business clients…once you have got one or two families in the community you get the whole lot.' She also recognized an increasing number of professional Asian women: 'Fifteen and sixteen year old kids…they are not really our clients, our clients are business people, we want the professionals.'

Malini's company now exports to other countries with significant Asian populations, notably South Africa and the USA through the establishment of a parallel export–import firm. She also has buyers who retail her products in Europe. 'We export from India to…Mauritius, South Africa and America, it's better to do direct rather than going by Great Britain because the duties are too high in Great Britain'. An American client undertakes quality checks and orders while travelling on business but as her parents also live in Great Britain such trips also become family visits. Malini thus actively uses diasporic connections to produce and sell her products but also she recreates a diasporic culture through production and marketing strategies. In Europe, Malini's business has begun to cater not just to the diaspora but also to the 'white population' for whom Indo-chic (Puwar, 2002) has become popular. In Britain Malini notes that she has 'regular customers who are English especially at Christmas time'.

The clothes she sells are designed in Great Britain and are influenced by British designers but manufactured in India. Malini feels that the skills required for the production of the clothes are only available in India: 'just [for] the labour [and] they produce what we want'. Most often, fashion advertising and marketing necessarily erase images of production, remembering only the processes of conceptualization and design. However, for diasporic fashions the place of production holds particular meaning. The consumption of the salwaar kameez is underpinned by diasporic links with South Asia, so that the salwaar kameez gains additional (marketable) authenticity because of its being produced in this space, the 'authentic' home of such garments. South Asia then becomes the site of production where the cultural product not only acquires value, but also a site where 'authenticity' is produced. Hence, the production history of Malini's fashion garments in India becomes something to be celebrated, rather than hidden. Production tales thus become central to consumption tales. All of this is not to deny the fact that Indian manufacturers provide lower production costs by way of cheaper labour, but rather to nuance existing stories about capital's search for 'low wage locations'.

After the Indian economy was liberalized in 1992 it became possible for Malini to open an office in India (the company is registered in the United Kingdom) and work more closely with subcontract cut, make and trim (CMT) units. She travels to India every two months to manage the business. Upgrading of garments to fashion products is dependent on aesthetic values of consumers which is discursively produced in the locales of consumption. Malini's input into the production of the clothes thus is vital. 'Authenticity' must be mediated to meet consumer choice, and this is difficult to achieve in differentiated production locales without frequent visits to monitor production.

In retail spaces, Malini only employs women. She argues that this practice is important even to male customers: they 'value a woman's opinion of what they look like rather than a guy's which is strange. We found that they prefer to be served by women, it's just worked out that it should be all women and why not?' Interestingly, although both Malini and her customers wear Western clothes on weekdays they wear Asian clothing on Saturdays. Retail centres thus can be viewed as spaces of consumption as well as performative spaces – spaces which not only reproduce but also create a sense of identity. Just as people 'go shopping' not merely to buy or even to see but also to be seen, so too those who sell clothes also dress in order to be seen. Differentiating between different days of the week also requires a sensitivity to who comes shopping and when. If Malini wears a salwaar kameez in her shop on Saturday, it is probably a recognition that the nature of the shopping area changes at the weekends, with members of the Asian community coming from further afield. The ethnic mix of the shopping area changes and 'dressing up' forms part of the creation of a milieu and the articulation of one version of group identity.

Understanding bodily performance in retail spaces is important to the analysis of consumption practices. As Leslie (2002, 74) reminds us, 'codes of performance' in most fashion retail outlets 'demand a white, middle-class script'. One of Leslie's (2002, 74) young interviewees recalled that 'when Indian people came in, she [the manager] would make comments about the way they smelled and dressed…I don't think she considered me Indian just because I didn't dress Indian or if you didn't…show your culture'.[10]

Selling Asian clothes requires showing culture and performing race in order to attract some customers – although at the same time this very performance may well limit access to other customers. The practice of dressing up in Asian clothes to go shopping/selling on Belgrave Road must also be seen in conjunction with exclusionary processes operating elsewhere, where wearing a salwaar kameez may attract racist comments such as 'Paki'. Just as people may dress in Asian clothes to go shopping in Wembley or on Belgrave Road, they also wear Western clothes to go shopping in the high street. Inclusionary and exclusionary processes are simultaneously juxtapositioned in the creation of specialist retail centres. It is important to recognize that the terms of participation within distinctive retailing centres such as Leicester's Belgrave Road are not wholly celebratory or autonomous but must be situated within the contingent contextuality of the spaces and places from which such participation is excluded. Although 'Asian' design elements have recently been incorporated into Western fashion, I would argue that this has not increased the 'acceptability' of Asian clothing per se. Moreover, public 'acceptability' is always mediated by the consuming body: Asian women wearing salwaar kameez may be cast as 'traditional' at the same time that these garments are being adopted and adapted by Western women as a mark of their modernity and cultural openness. As Malini's experiences demonstrate, producing and wearing clothes requires a sensitivity not simply to ethnicity or to class and gender, but also to spatially situated racialized codes.

Conclusions

The central aim of this chapter has been to highlight the agency of South Asian women in the production of fashion garments in the diaspora. Through the story of Malini, I have documented the emergence of a specialized niche within the fashion commodity chain. The development of this niche owes much to the growth of a significant diasporic social class with high purchasing power, but I have also sought to emphasize the specific role of South Asian women in commodity chain initiation. Further, I have suggested that in seeking to fill this niche, women such as Malini have drawn upon their own consumption patterns and choices. The chapter has also demonstrated that a consideration of the personal biographies of commodity chain actors alongside product biographies can contextualize and deepen understandings of the histories and spatialities of consumption and production. This method has been used not only to provide an orienting map of the fashion commodity chain but also to challenge existing orientalizing narratives of the agency of South Asian women.

Notes

1 See also Dolan and Tewari's (2001) comparison of the food and clothing sectors.
2 Lash's own work in fact adopts the product biography as the primary method – as the 'perspective or orientation within which a range of appropriate methods is deployed' (1997, 6) – which ultimately objectifies the production–consumption chain.
3 I use the term Third World women to refer to women who face commonalities in their struggles against imperialism, racism and sexism (Mohanty, 1991). It is also important to draw attention to the role of colonial legacies and imperialist narratives in constructing representations of such women – whether 'placed' in the Third World or dis/placed in the First through migration. These legacies continue to condition the ways in which women are read in production and consumption narratives of the global political economy.
4 Occasional trips to the subcontinent presented another opportunity to 'stock up' on clothes. Although such trips are now more affordable and are therefore becoming more common, the need to 'stock up' has on the other hand reduced.
5 Although the small business literature might offer space for such discussion, this body of work has retained a strongly empirical focus, concerning itself with policy issues such as identifying appropriate mechanisms for supporting businesses.
6 The larger research project from which the case study derives was funded by The Nottingham Trent University. My colleagues on this project were Irene Hardill, David Graham and Adam Strange. This section draws in particular upon a three and a half hour interview with Malini (a pseudonym) conducted by me in 1996. However, the broader argument has mainly been influenced by conversations with colleagues on other projects, particularly with Nirmal Puwar, Estella Tincknell, Richard Johnson and Deborah Chambers. The views expressed here are mine alone.
7 At the same time it is important to emphasize that unemployment rates remain much higher for minority ethnic populations (8.3 per cent for whites compared to 17.6 per cent for all minority ethnic groups). These figures vary between

groups, ranging from 12.6 per cent among Indians to 25.8 per cent among Pakistani-Bangladeshi populations.

8 The emergence of a South Asian minority with considerable purchasing power is not perhaps wholly new. I have argued elsewhere (Raghuram, 2000) that women have played an important part in skilled migration to the United Kingdom but that their presence has not generally been recognized in migration literature. What is new perhaps is the self-confidence of a generation of professionals who have been brought up in the United Kingdom.

9 Clothing such as lehengas have also become popular although the popularity of these garments waxes and wanes.

10 Problematically, however, Leslie (2002, p. 74) argues that 'performances require employees not to foreground their ethnic or racial identities'. Despite her earlier recognition of the white middle-classness of this script, she then proceeds to universalize white identity as the norm, as lacking an ethnic identity, as an empty sign. As Dyer (1997) has suggested, this emptiness is a sign of racial privilege reserved for 'white bodies' so that only non-white bodies are seen as performing ethnicity.

References

Appadurai, A. (ed.) (1986) *The Social Life of Things: Commodities in Cultural Perspective*. Cambridge: Cambridge University Press.

Bhachu, P. (1995) New cultural forms and transnational Asian women: culture, class and consumption among British Asian women in the diaspora. In van de Vere, P. (ed.), *Nation and Migration: The Politics of Space in the South Asian Diaspora*. Philadelphia: University of Pennsylvania Press, pp. 56–85.

Bhachu, P. (2000) Dangerous designs: South Asian fashion and style in global markets. Public Lecture for ESRC Transnational Communities Project, University College London.

Bourdieu, P. (1984) *Distinction: A Social Critique of the Judgement of Taste*. Cambridge, MA: Harvard University Press.

Bujra, P. (1992) Ethnicity and class: the case of the East African Asians. In Allen, T. and Thomas, A. (eds), *Poverty and Development in the 1990s*. Milton Keynes: Open University, pp. 437–61.

Byrne, D. (1998) Class and ethnicity in complex cities: Leicester and Bradford. *Environment and Planning A*, 30, pp. 703–20.

Callon, M. (1991) Techno-economic networks and irreversibility in a sociology of monsters'. In Law, J. (ed.), *A Sociology of Monsters: Essays on Power, Technology and Domination*. London: Routledge, pp. 132–61.

Cook, I., Crang, P. and Thorpe, M. (2000) Regions to be cheerful: culinary authenticity and its geographies. In Cook, I., Crouch, D., Naylor, S. and Ryan, J. (eds), *Cultural Turns/Geographical Turns*. London: Pearson Education, pp. 109–39.

Craik, J. (1994) *The Face of Fashion: Cultural Studies in Fashion*. London: Routledge.

Crewe, L. and Davenport, E. (1992) The puppet show: changing buyer supplier relationships within clothing retailing. *Transactions of the Institute of British Geographers*, 17, pp. 183–97.

Dicken, P. and Hassler, M. (2000) Organizing the Indonesian clothing industry in the global economy: the role of business networks. *Environment and Planning A*, 32(2), pp. 263–80.

Dicken, P., Kelly, P., Olds, K. and Yeung, H. (2001) Chains and networks, territories and scales: towards a relational framework for analysing the global economy. *Global Networks*, 1(2), pp. 89–112.

Dolan, C. and Tewari, M. (2001) From what we wear to what we eat: upgrading in global value chains. *IDS Bulletin*, 32(3), pp. 94–104.

Du Gay, P., Hall, S., Janes, L., Mackay, H. and Negus, K. (1997) *Doing Cultural Studies: The Story of the Sony Walkman*. London: Sage.

Dyer, R. (1997) *Whiteness*. London: Routledge.

Gereffi, G. (1994) The organisation of buyer-driven global commodity chains: how US retailers shape overseas production networks. In Gereffi, G. and Korzeniewicz, M. (eds), *Commodity Chains and Global Capitalism*. Westport: Praeger, pp. 95–122.

Gereffi, G. (1999) International trade and industrial upgrading in the apparel commodity chain. *Journal of International Economics*, 48(1), pp. 37–70.

Gibbon, P. (1997) Prawns and piranhas: the political economy of a Tanzanian private sector market chain. *Journal of Peasant Studies*, 24(4), pp. 1–86.

Gregson, N., Brooks, K. and Crewe, L. (2000) Narratives of consumption and the body in the space of the charity shop. In Jackson, P., Lowe, M., Miller, D. and Mort, F. (eds), *Commercial Cultures, Economies, Practices and Spaces*. Oxford: Berg.

Gwynne, R. (1999) Globalization, commodity chains and fruit exporting regions in Chile. *Tijdschrift voor economisce en social geografie*, 90(2), pp. 211–25.

Hartwick, E. (2000) Towards a geographical politics of consumption. *Environment and Planning A*, 32, pp. 1177–92.

Hills, J. (1998) *Income and Wealth: The Latest Evidence*. York: Joseph Rowntree Foundation.

Holmquist, C. and Sundin, E. (1989) The growth of women's entrepreneurship: push or pull factors? Paper presented to EISAM Conference on Small Business, University of Durham Business School.

Hughes, A. (2000) Retailers, knowledges and changing commodity networks: the case of the cut flower trade. *Geoforum*, 31(2), pp. 175–90.

Jackson, P. (1999) Commodity cultures: the traffic in things. *Transactions of the Institute of British Geographers*, 24, pp. 95–108.

Johnson, R., (1996) The story so far: and further transformations? In Punter, D. (ed.), *Introduction to Contemporary Cultural Studies*. London: Longman, pp. 277–313.

Johnson, R., Chambers, D., Raghuram, P. and Tincknell, E. (2004) *The Practice of Cultural Studies*. London: Sage.

Lash, S. (1997) *The culture industries: Biographies of cultural products*. Available at http://www.goldsmiths.ac.uk/cultural-studies/html/cultural.html

Latour, B. (1993) *We Have Never Been Modern*. Hemel Hempstead: Harvester Wheatsheaf.

Leslie, D. (2002) Gender and fashion retailing employment. *Gender, Place and Culture*, pp. 61–76.

Leslie, D. and Reimer, S. (1999) Spatializing commodity chains. *Progress in Human Geography*, 23, pp. 3401–20.

Mani, B. (2002) Undressing the Diaspora. In Puwar, N. and Raghuram, P. (eds), *South Asian Women in the Diaspora*. Oxford: Berg.

Marlow, S. and Strange, A. (1994) Female entrepreneurs: success by whose standards? In Tandton, M. (ed.), *Women in Management: A Developing Presence*. London: Routledge, pp. 172–84.

Mather, C. (1999) Agro-commodity chains, market power and territory: re-regulating South African citrus exports in the 1990s. *Geoforum*, 30(1), pp. 61–70.

Metcalf, H., Modood, T. and Virdee, S. (1996) *Asian Self-employment: The Interaction of Culture and Economics in England*. London: Policy Studies Institute.

Mies, M. (1982) *The Lacemakers of Narsapur, Indian House-wives Produce for the World Market*. London: Zed Press.

Mitchell, K. (1995) Flexible circulation in the Pacific rim: capitalisms in cultural context. *Economic Geography*, Oct 71, 4, pp. 364–82.

Mohanty, C. (1991) Cartographies of struggle: Third World women and the politics of feminism. In Mohanty, C., Russo, A. and Torres, L. (eds), *Third World Women and the Politics of Feminism*. Bloomington: Indiana University Press, pp. 1–50.

Mohanty, C. (1997) Women workers and capitalist scripts: ideologies of domination, common interests, and the politics of solidarity. In Alexander, J. and Mohanty, C. (eds), *Feminist Genealogies, Colonial Legacies, Democratic Futures*. London: Routledge, pp. 3–30.

Munshi, S. (2001) Marvellous me: the beauty industry and the construction of the modern Indian woman. In Munshi, S. (ed.), *Images of the 'Modern Woman' in Asia: Global Media, Local Meanings*. Richmond: Curzon, pp. 78–93.

Nash, J. and Fernandez-Kelly, M. P. (1983) *Women, Men and the International Division of Labour*. Albany: SUNY Press.

Nash, D. and Reeder, D. (1993) *Leicester in the Twentieth Century*. Stroud: Alan Sutton Publishing in association with Leicester City Council.

Nava, M. (1991) Consumerism reconsidered: buying and power. *Cultural Studies*, 5, pp. 157–73.

O'Connor, H. (1995) *The Spatial Distribution of Ethnic Minority Communities in Leicester, 1971, 1981, and 1991: Analysis and Interpretation*. Leicester: Joint Publication of the Centre for Urban History and the Ethnicity Research Centre, University of Leicester.

Olds, K. and Yeung, H. W. (1999) (Re)shaping 'Chinese' business networks in a globalising era. *Environment and Planning D*, 17, pp. 535–56.

Ong, A. (1987) *Spirits of Resistance and Capitalist Discipline: Factory Women in Malaysia*. Albany: SUNY Press.

Owen, D. (1995) *Ethnic Minority Women and Employment*. Manchester: Equal Opportunities Commission.

Phizacklea, A. and Ram, M. (1996) Being your own boss: ethnic minority entrepreneurs in comparative perspective. *Work, Employment and Society*, 10, pp. 319–39.

Puwar, N. (2002) Multicultural fashion…stirrings of another sense of aesthetics and memory. *Feminist Review*, 71, pp. 63–87.

Raghuram, P. (2000) Gendering skilled migratory streams: implications for conceptualising migration. *Asian and Pacific Migration Journal*, 9, pp. 429–57.

Raghuram, P. and Hardill, I. (1998) Negotiating a market: a case study of an Asian woman in business. *Women's Studies International Forum*, 21, pp. 475–84.

Raikes, P., Jensen, M. F. and Ponte, S. (2000) Global commodity chain analysis and the French filière approach. *Economy and Society*, 29, pp. 390–417.

Tarlo, E. (1996) *Clothing Matters: Dress and Identity in India*. London: Hurst and Co.

Van-Grunsven, L. and Smakman, F. (2001) Competitive adjustment and advancement in global commodity chains: firm strategies and trajectories in the East Asian apparel industry. *Singapore Journal of Tropical Geography*, 22, pp. 173–88.

Yeung, H. W.-C. (2000) Embedding foreign affiliates in transnational business networks: the case of Hongkong firms in Southeast Asia. *Environment and Planning A*, 32, pp. 201–22.

Part III

Commodities, representations and the politics of the producer–consumer relation

7 Geographical knowledges in the Ecuadorian flower industry

Justine Coulson

Introduction

There has been much analysis of the ways in which commodities are situated in geographical representations and performances as they flow through commodity chains (Cook *et al.*, 2000; Crang, 1996; Jackson and Taylor, 1996). By tying a product to a specific geographical region or cultural identity, producers aim to entice consumers with promises of authenticity and/or 'exoticism'. Drawing upon a broader study of female flower workers in Ecuador (Coulson, 2000), this chapter critically examines advertising images used in promotional literature for the national flower industry. In particular, I focus upon the ways in which Ecuadorian flower companies have drawn upon images of Ecuadorian femininities and ethnicities in order to enhance the geographical knowledges that surround the product and to provide it with an identity. Companies create different 'geographical lores' (Crang, 1996) for their product and as a result draw upon diverse aspects of Ecuadorian femininities and ethnicities. In all three advertisements considered in the chapter, a signifying relationship is created between the images of the landscape, the person and the product. Yet in each of them the representation of Ecuadorian femininity and ethnicity within this triangular relationship plays a different role.

Geographical knowledges and the depiction of people in advertising

The creation of geographical knowledges involves an ever-shifting relationship between the depiction of places and the depiction of people. Landscapes and bodies may work together to create an identity for a product. In some instances, the bodies in an image may be inscribed with geographical signifiers in such a way that the depiction of location is no longer necessary. Alternatively, the body serves a limited role in creating geographical knowledge for a product but takes on those qualities suggested by the landscape in which it is situated. People appear as representations of producers, consumers and abstract values, and these roles are not mutually exclusive.

Brandth and Haugen's (2000) analysis of advertising images in the Norwegian forestry press suggests that images of people serve as a medium through which to channel the qualities suggested by the landscapes to the potential consumer. The harsh frozen landscapes filled with rocks, thunder and lightening created through text and images serve not only to evoke qualities of power, ruggedness and reliability in the logging machinery advertised, but they also infuse the images of men that appear in the advertisements with a hyper-masculinity. The represented body serves as a link between the qualities of the product and the qualities the consumer desires to identify in himself, and the geographical images of a forestry scene serve to emphasize those qualities. As products often are aimed at either a male or female consumer, representations of consumers are deliberately gendered and the landscapes within which they are located may also be gendered. Just as the macho Norwegian lumberjack is depicted pitted against the forces of nature in a battle of strength, the female consumer is most commonly depicted in a domestic setting as wife and mother (Leslie, 1993).

When indigenous and non-white people are used in advertising images to depict consumers, they may be associated with creating inclusive, diverse geographical imaginaries or broad global landscapes. Such images, when targeted at the white consumer create a uniqueness for their product through the representation of an 'exotic' other (O'Barr, 1994). However, when aimed at an ethnically diverse consumer group, the intention may be to establish the broad appeal of a product, as in the case of Pepsi and Coca-Cola marketing strategies in Papua New Guinea (Foster, 1999). Foster's analysis illustrates the way in which the two cola companies draw on images of ethnic diversity not to associate their products with an 'exotic other', but to create the concept of an ethnically harmonious, inclusive nation-state. The consumption of a single brand is then linked to this image of a unified but diverse nation. Advertisements presented by Foster (1999) show that both Pepsi and Coca-Cola have relied on images of people to create an imaginary Papua New Guinea rather than images of specific landmarks. Backdrops suggest sun and lush vegetation, which could be true of any tropical island in the Pacific region. While the Pepsi and Coca Cola advertisements specifically locate the images of consumers in an approximation of a Papua New Guinean landscape, no matter how featureless, Benetton's advertisements have often offered images of people/consumers that are set against plain white backdrops, and thus, are devoid of references to physical landscapes (Jackson and Taylor, 1996). In order to create a global brand that is promoted on the grounds of racial and cultural harmony, images of ethnically diverse people seek to invoke a 'global village' constituted by consumers of Benetton products (ibid.). Consumption takes place anywhere and everywhere, and therefore one can project one's own landscape onto the location-neutral background.

As well as offering images of consumers that allow the actual consumer to imagine herself or himself as a new person or as a member of a wider community through consumption, advertisements that create geographical knowledges for their product may also depict idealized landscapes that are devoid of people. Crang's (1996) analysis of the 'geographical lores' foregrounded in a UK retailer's bath products reflects upon this creation of unpopulated 'landscapes'. A sweeping rhetoric that creates geographical images for each of the cardinal points sets up a direct relationship between the product, the consumer and the imaginary landscape. It is the consumer alone who will be transported to 'steamy jungles', 'an island paradise' and 'the freshness of Arctic snows' (Crang, 1996, 53). In constructing bathing as a personal moment of self-indulgence, there is no space within these advertisements for anyone else. Similarly, advertising images used by the South African company Outspan seek to create a generic African landscape (Mather and Rowcroft, Chapter 8, this volume). Portrayals of consumption by animals rather than humans reinforce the idea of a wholly 'natural' product grown in the stereotypical 'wilds of Africa'.

While the depiction of consumers in the aforementioned examples can be linked to two forms of geographical knowledge identified by Crang (1996) – namely, broad global displacements and the representations of sites of consumption and product use – the depiction of producers in advertisements can be linked to what Crang identifies as 'tightly specified and realist knowledges about commodity production' (1996, 54). Such representations of producers are linked to creating messages of authenticity, tradition and exoticism, which are further developed through the depiction of 'foreign' landscapes. Images of Guatemalan handicraft producers in US mail-order catalogues also tend to use simple representations of the product and the producer in Guatemalan landscapes in order to convey a 'genuine' product to American consumers (Hendrickson, 1996). Images which foreground the manufacture of the product by hand underscore textual messages of uniqueness and tradition (ibid.). While images of producers and sites of production can work together to create an 'authentic' identity for a product, images of producers can stand alone in the creation of geographical knowledges. In Cook *et al.*'s (2000) analysis of the industrial mass production of 'ethnic' foodstuffs, products often are presented as authentic through narratives of the origins of the people involved in production.

When using images of people to create a geographical identity for a product, it is not necessary for such images to depict only consumers and producers. Bodies are used in advertising to represent or embody abstract values, and these abstract values can form part of the geographical 'lores' surrounding a product. For example, the Outspan girls discussed by Mather and Rowcroft (Chapter 8, this volume) represent neither consumers nor producers. Bronzed, slim bodies that are instructed to skip through the crowds create an association between the product and the qualities of fun and health in the mind of the consumer. As white Afrikaner women instructed not to discuss the political situation in South Africa, they are the embodiment of the depoliticized, non-black imaginary South Africa that Outspan wished to create for its product. In the use of bodies to depict abstract values, the body may become anonymous and as much an object for consumption as the product. The non-white and the female body are regularly dissected and fetishized in advertising images (Bordo, 1993; Nayak, 1997).

Selling Ecuadorian cut flowers to the world: the question of product identity

Before considering the representation of Ecuadorian femininities in promotional literature for the national flower industry, it is necessary to provide some background on the development of commercial floriculture in Ecuador. The first cut flower plantation was established in 1983, but production accelerated during the first half of the 1990s. In 1989 the value of annual flower exports stood at US$2.5 million and by 1998 had risen to US$141 million (Mena, 1999). Cut flower production is hailed nationally as an exemplar of successful non-traditional export development, and while the World Bank has criticized Ecuador in general for failing to develop an export mentality and capacity, the flower industry has been singled out as the one example of 'successful entrepreneurship' (World Bank, 1995). The industry is characterized by majority national ownership: in 1996, 80 per cent of all flower companies were owned by Ecuadorian nationals (Colitt, 1996). Exports to the United States, Europe and the Russian Federation account for 94 per cent of the flowers exported. There are three main patterns of marketing: selling directly to importers (especially for flowers exported to the United States, Canada and Europe); selling through a contract with a specific retailer; and selling through exporters based in Ecuador (mainly exports to the Russian Federation) (Palán and Palán, 1999).

The slow development of the industry in the 1980s can be attributed to a number of factors, such as inadequate air cargo capacity, inept cargo handling and insufficient technical expertise to produce high-grade flowers. The Ecuadorian Association of Flower Producers and Exporters (EXPOFLORES) responded to this challenge by lobbying assiduously for better infrastructure and economic assistance for commercial flower growers, by importing floriculture specialists from Colombia and by providing training and technical assistance programmes. The industry also benefited from such comparative advantages as relatively low land, water and labour costs in comparison to other Latin American flower producers and a climate that permits year-round production (C.F.N., 1995).

A key problem facing the Ecuadorian flower industry during the 1980s and early 1990s – and one which is central to my discussion in this chapter – is the issue of product identity. Even if a flower company can offer high quality, modern varieties at a competitive price, this may not be enough to ensure a market for their product. Competition in international cut flower production is increasing as the industry continues to attract new producers, such as India and China (van Liemt, 1999). Therefore, it is inevitable that other producers in the international market will be offering the same variety with equivalent quality at an equally competitive price. This intensification of competition makes successful marketing of the utmost importance. However, up until the mid-1990s, Ecuadorian flowers as a relatively unknown, untested newcomer to the market were not associated with outstanding quality in the mind of the European or North American intermediate buyer.

Ecuador's proximity to one of the world's largest flower exporters, Colombia, was initially a disadvantage in terms of launching Ecuadorian flowers onto the international market. Having been established some fifteen years before the Ecuadorian industry, the international reputation of Colombian floriculture was already well developed when Ecuador first began to produce flowers (Meier, 1999). As a much smaller, lower-income country, Ecuador was overshadowed by a middle-income neighbour that could boast a higher international industrial profile. EXPOFLORES dealt with this problem by establishing the Ecuadorian source of the product as the equivalent of a brand name that would suggest high quality flowers.

Brands infuse products with a personality, leading consumers to believe that individual brands are of superior quality and that they will positively impact upon their lifestyle (De Chernatony and McDonald, 1997). In a consumer world where there is very often very little difference between similar goods from different companies, brands are essential to convincing consumers that they are buying a unique product (Klein, 2000, 20). In most cases, a brand travels physically with a product to the point of end consumption, whether a Nike symbol sewn onto a sweatshirt or a Chiquita label stuck on a banana. However, branding in the case of Ecuadorian flowers is somewhat different. As the majority of flower companies in Ecuador sell to wholesale exporters and importers, rather than directly to the consumer, any brand created for Ecuadorian flowers is an ephemeral one. If buyers can be encouraged to purchase flowers from EXPOFLORES members because they associate Ecuador with high-quality flowers, then the brand creation has served its purpose. At the point of sale the flower may take on a new brand identity of the retailer or may be sold as an anonymous product from a bucket of flowers in a market stall. In order to ensure that Ecuador as a product identity became synonymous with high quality flowers in the minds of intermediary buyers, EXPOFLORES strove to ensure that Ecuadorian flowers became a visible, identifiable product at an international level. The organization financed stands in international trade fairs, and significantly, although representatives from only a few of the best plantations were sent, they displayed their flowers under a group identity as Ecuadorian producers rather than as individual companies.[1] The approach was highly successful: having won a number of gold medals at international trade fairs such as that held in Aalsmeer, the Netherlands, Ecuadorian flowers and particularly roses, came to be considered to be among the best cut flowers in the world (Haines, 1994).

Although national producers are represented by EXPOFLORES at an international level, individual companies must market and sell their own flowers. In order to attract the attention of buyers, many individual companies have chosen to promote their product based on the strength of the national brand identity created by EXPOFLORES. However, although a series of individual companies may all evoke images of Ecuador as a means of marketing their product, by drawing on diverse images they create different versions or 'lores' of Ecuador to promote their product.

Resolving issues of product identity: advertising images in trade promotion literature

In order to explore how Ecuadorian flower producers have used advertising to create a product identity, three advertisements that appeared in a 1996 floriculture trade show brochure will be discussed. This annual trade fair, held in Quito, allows producers based in Ecuador to meet buyers and service providers located outside the country. The accompanying brochure (which lists all participants) serves as a trade directory and therefore placing an advertisement in this publication enables producers to attract the attention of buyers. The majority of the colour full-page and half-page advertisements that appear in the brochure represent foreign companies that provide services such as specialist cargo facilities and imported irrigation systems. The three advertisements analysed here are distinctive: they are the only colour advertisements representing national flower growing companies in the brochure that go beyond a simplistic representation of the product. Other advertisements in the trade fair catalogue contain nothing more than the name of the company, a photograph or graphic representation of a single flower and in some cases a slogan. However in the three advertisements discussed here we see examples that use a more complex montage of images that aim to infuse the product with certain meanings and thus create an identity for the product.

The aim of any advertisement is to capture a larger market share for a product by making it more desirable to consumers. This is achieved by taking the product and images out of their normal contexts and placing them together within the frame of the advertisement. Images, texts and meanings come together to create a 'commodity-signifier' that gives the product a new meaning (Goldman, 1992). My central concern in analysing these images is twofold: to understand what meanings the following advertisements establish for the cut flowers they are promoting; and to consider how images of people are used to create those meanings.

All three advertisements focus on the site of cultivation (i.e. Ecuador) as the source of identity for their product. By claiming an Ecuadorian identity for their product, they build on the work done by EXPOFLORES to promote the Ecuadorian flower industry. However, although all three companies claim a national identity for their product, through the compilation of different images and text, the significance of the product's origins is distinct in each advertisement. The first advertisement offers images of a modern, developed Ecuador that suggests that the flowers produced there are the product of a modern growing process. In the second advertisement the relationship established between Ecuador and the product is not one of national characteristics being conveyed to the product, but of the product giving something to Ecuador. Here, we are presented with a rural, indigenous version of Ecuador, and flower production is linked to community development. The third advertisement offers us neither symbolic representations of an urban nor a rural Ecuador. Rather than creating geographical knowledges to sell the product outside Ecuador, the advertisement is aimed at selling consulting services to Ecuadorian growers. As a result, the version of Ecuador presented is that of the actual site of production – namely, a greenhouse where the flower is being sorted.

In each of the advertisements, the 'geographical lores' constructed around the product are formed at least in part by images of people. The bodies presented in the advertising images are inscribed with both gender and specific ethnicities in order to inform the Ecuadorian identities created for the products presented. An analysis of these portrayals of ethnicity and gender shows that while responding to the perceived interests and values of potential buyers, the advertisements are also shaped by particular discourses of ethnicity and gender found in Ecuador.

Ecuador as a 'green' location

LoveRoses is a company that sells roses to wholesale importers. The LoveRoses advertisement (Plate 7.1) draws on a particular set of images of Ecuador that imbue the product with certain qualities which aim to convince a potential wholesaler of the quality of the product. The advertisement establishes a national identity for the flowers by referring to a compounded geography and history that renders the country unique. Ecuador's equatorial location in itself is not exclusive. However, Ecuador's role in the history of the measurement of the equator

Plate 7.1 LoveRoses advertisement.

Source: Agriflor de las Americas catalogue, 1996.

is enshrined in the country's name, and has become a key element in discourses of national identity (Radcliffe and Westwood, 1996). The advertisement makes reference to this definitive aspect of Ecuadorian identity through the use of an image of the '*Mitad del Mundo*' (Middle of the World) site, a museum complex that tells the story of the scientific community's attempts to identify the position of the equator. The obelisk that lies at the centre of the advertising image is situated on the equatorial line, and marks the point at which the equator was first successfully measured in the nineteenth century. By selecting the equatorial location of Ecuador as the aspect of national identity with which to brand its product, the company has chosen an image that suggests both national heritage and scientific precision. The decision to project these specific qualities onto the product almost certainly derives from the national industry's concerns about the reputation of Ecuadorian flowers in the national market. First, in response to the overshadowing of the Ecuadorian flower industry by its Colombian counterpart, the advertisement's claim to a particular history suggests an established nation with a distinct identity that sets it apart from its Andean neighbour. Second, the allusion to scientific precision responds to any continuing worries by buyers outside Ecuador that the industry encounters problems with ensuring uniform quality. The quality of Ecuadorian flowers was an issue in the early 1980s, but following a USAID-funded technical assistance programme and an influx of Colombian specialists, standards have improved considerably, and in the 1990s the Ecuadorian flower industry won several gold medals at international exhibitions (Haines, 1994).

Having established the geographical identity of the product and inferred positive characteristics associated with that identity, the slogan in the advertisement then draws upon the 'nationality' of the flowers to explicitly claim a product-enhancing quality. The text links the concept of a superlatively 'green' product ('the cleanest roses') with the geographical location in which they are grown ('the middle of the world'). The meaning of the term 'clean' is ambiguous: it could suggest either that the flowers have been grown without the intensive application of pesticides, or alternatively, that the flowers would pass strict fitosanitary tests and therefore, they had been grown with a more intensive use of chemicals. 'Clean' could be read as a reflection that production is non- or less polluting or it could be seen to guarantee the complete absence of any living insect or mite. Through such ambiguity the advertisement responds to conflicting demands at different sites of consumption. While flower imports into the United States and Japan must pass rigorous fitosanitary standards that often result in flower producers using chemicals to prevent rather than react to diseases and infestations (Maharaj and Dorren, 1995), in Europe there is increasing pressure from environmental groups and retailers to cultivate flowers in a more environmentally sustainable way and therefore to reduce the amount of chemicals used (van Liemt, 1999).

While the meanings suggested by the background image and the text might seem relatively straightforward, the model in the advertisement has a less clearly defined role in relation to the promotion of the flowers. She is holding a bunch of roses that have been arranged in a bouquet and therefore, prepared for consumption. Her dress, jewellery and hairstyle also suggest an urban femininity rather than a rural flower worker. Thus it appears that she is not to be read as a producer/worker. However although she is holding 'consumed' flowers, she cannot represent a consumer. First-grade roses are grown only for export, and the background does not represent the site of consumption, but rather the site of production. To see her as a consumer would be to suggest the product were only suitable for the domestic market, which is contrary to the message the company wishes to convey about its product through the advertising image.

Through her relatively ambiguous location – that is, standing 'outside' the production/consumption chain – the female model comes to play a more abstract role as part of the commodity-sign that defines the Ecuadorian identity and quality of the product. At one level she can be read as an embodiment of the qualities of the product. Among predominantly male flower professionals in Ecuador, analogies constantly are drawn between the perceived qualities of cut flowers and those of the women, be they women who work among the flowers, who model flowers or who receive them as gifts. Both women and flowers are referred to as 'beautiful', 'fragile' and 'liking attention', and flower professionals recreate the connection between women and nature that has been witnessed in other non-traditional horticultural export workplaces in Latin America (Berrecil, 1995, 186). This relationship is recreated visually in the advertisement where the female body serves as another element of the background that creates an identity for the flowers. By framing the images of woman and flowers within the advertisement both become the passive objects of what Rose (1993, 97) describes as an eroticising, heterosexual masculine gaze, which in this case is the predominantly male world of flower production and trading.

However, as well as underscoring the 'feminine' qualities of the flower, one could also read the image of the model as the embodiment of a *mestizo*[2] national identity. Representations of woman and nature are often infused with messages regarding race and ethnicity (de Oliver, 1997), and this image is no exception. There continues to exist in Ecuador a discourse of an all-inclusive *mestizo* nationhood that seeks to erase indigenous identity as an embodied practice. This is promulgated through political, media and educational channels (Radcliffe and Westwood, 1996), and represents the vestiges of official attempts in the 1970s and 1980s to create a non-ethnic Ecuadorian identity, and therefore, an undivided nation-state (Stutzman, 1981). Within this discourse, *mestizo*/white identities are associated with what are considered to be positive qualities, such as 'progress', 'modernity' and 'hygiene',[3] while indigenous identities are associated with the inverse values (Almeida, 1996, 55). Practices within the flower industry reveal the adherence by many flower professionals to these values (Coulson, 2000). Therefore, in creating an Ecuadorian 'lore' for the product that suggests precise and sanitary production, the use of a non-European/North American, non-indigenous, and hence, by default *mestizo* body in the advertising image serves to underscore this message of a 'developed' Ecuador, while the gender of the body brings more general values associated with femininity to the product.

Ecuador as a land of indigenous peoples

While the advertisement for LoveRoses (Plate 7.1) projects the message that the unique geographical location of Ecuador makes its product superior, the advertisement for Malima (Plate 7.2), also aimed at wholesalers, argues that its product gives something back to Ecuador through community development. This is suggested in the slogan that appears in the text ('When a flower grows at Malima, those who have taken care of it also grow healthy'), and the additional text underlines the company's commitment to labour and environmental standards. This is an explicit response to pressure from national NGOs (Mena, 1999), international pressure groups such as the Food Information and Action Network (FIAN), and retailers such as the Swiss chain Migros (van Liemt, 1999), for better health and safety controls and general conditions for flower workers. While the meaning of the text is clear, it is given additional significance by the montage of images that surround it.

There are four key elements to the commodity-sign created by this advertisement other than the text: the flowers that form the background, the image of a smiling woman among the foliage, the child, and the central image of three flowers. All the images work together to underline the message that by buying this product, one is supporting the welfare of the local population.

In the other advertisements, a photographic image of the product appears. In the Malima advertisement, however, the flowers at the centre of the advertisement are not real. They are a simplistic representation of flowers cut from paper, in bold, unnatural colours. A picture of the actual product is not necessary: the advertisement aims to foreground the social development benefits achieved by buying Malima flowers rather than product quality. The image suggests a child's handiwork and therefore, serves to link the company's flowers with children.

Similarly, other flowers that form the background to the advertisement do not also serve to illustrate the quality of the product. In contrast to the image of *Mitad del Mundo* (Plate 7.1) which suggests national heritage and scientific precision, the background in the Malima advertisement is disorderly. The flowers are all green, and this colour, coupled with the abundance and lack of order, suggest fecundity and nature. The image has a double significance. The message relayed is that Malima flowers grow well and are natural, but the image also is linked to the connection between the growth of flowers and the growth of children. If Malima's flowers grow so abundantly, so too will the children of the workers.

As the link between flower production and family welfare is the key message of the advertisement, the image of the child beside the flower comes as no surprise. The use of children in advertisements using a camera angle that points slightly downwards is a standard method for linking a product with the value of nurturance (Messaris, 1997, 40). The other element of the advertisement is the image of a woman tending plants. She is half hidden by foliage and text. The theme of the advertisement causes us to assume that this woman is the child's mother. Just as she is seen happily nurturing the plants, she also nurtures her child and the flower company helps her to do so successfully.

Plate 7.2 Malima advertisement.

Source: Agriflor de las Americas catalogue, 1996.

In considering the images of Ecuador upon which the company has chosen to draw, it is necessary to reflect more closely upon the representation of the child in the image. Her non-Western dress style is evocative of an indigenous identity, and is clearly distinct from the formal dress of the model in Plate 7.1 and the uniforms of the women pictured in Plate 7.3. However, rather than representing actual communities from the region in which the plantation is located, the child appears as a composition, or an icon of indigeneity. She is wearing a hat that bears no resemblance to anything worn by any indigenous groups within Ecuador or in the Andean region in general. Arguably, the texture of the hat bears most resemblance to the texture of the paper flowers, and possibly the intention is to make the link between growing flowers and growing children more explicit. Her hair is curly and tied in two plaits. While the style of two plaits is associated with indigenous dress styles within the Andean region, the hair is traditionally heavily greased, with a clear middle parting and kept straight and close to the head. The white blouse is redolent of the type of embroidered blouse worn by women from the Otovalo region of Ecuador. However, as Otovalo is more than twelve-hour bus journey away from the region in which this company operates, it is a style unlikely to be seen in the plantation area. Her shawl appears to be more in the style of ikat or maybe, Guatemalan textiles. The few women living in communities close to the company's plantation who continue to wear what is identified as 'traditional' dress, normally wear a knitted, woollen shawl, while indigenous women from the Otovalo region wear a plain black shawl.

The image of the child represents a compromise that is made possible by the non-Ecuadorian identity of the targeted wholesaler. For many rural women working in Ecuadorian flower plantations, the experience of the workplace has encouraged them to adopt a more Western dress style, while it has been suggested that women who continue to wear 'traditional' clothes associated with an indigenous identity are treated badly by many plantation managers (Coulson, 2000). The indigenous sector of Ecuadorian society is viewed by the elite as a 'backward' presence that prevents the nation from achieving economic development (Radcliffe and Westwood, 1996). Unsurprisingly, this prejudice influences the attitude of some managers towards plantation workers. However, although the national elite may not value the indigenous population, there is a perception, based on the number of tourists that flock to the Otovalo handicraft market and images appearing on postcards, that Europeans and North Americans do (Korovkin, 1998). This advertisement responds to consumer expectations without identifying the producer too closely with an Ecuadorian indigenous identity. Furthermore, it reflects the paradoxical relationship between a *mestizo* rejection of modern-day indigenous identities while simultaneously evoking indigenous ancestors in order to create a national uniqueness (Muratorio, 1994; cited in Radcliffe, 1999, 38). Images of indigenous Ecuadorians can be a means of creating an 'authentic' or an 'exotic' identity for a product sold to non-nationals. There is no need, however, for that image to seek to portray actually existing 'reality',[4] as the image in isolation is unimportant. The image of the child forms part of the commodity-signifier, and it is assumed that the consumer will be able to 'read' the message conveyed by the entire advertisement.

Ecuador as flower plantation

Plate 7.3 is an advertisement for a flower growing company based in Ecuador that also offers a consultancy service to other flower producers. It is aimed at smaller scale Ecuadorian flower producers who are unable to afford full-time technical assistance.

Plate 7.3 Flortec advertisement.

Source: Agriflor de las Americas catalogue, 1996.

While the first two advertisements are concerned with creating an image of Ecuador with which to 'brand' their product for an international wholesaler, this advertisement seeks to convince national producers of the company's experience of Ecuadorian flower growing. While the text underlines the company's international expertise and professionalism, the image depicts a scene that would be familiar to any national producer and could have been taken in any one of the smaller plantations.

Unlike montage of images used in the first two advertisements, which build up an identity for the product, this advertisement uses a single photograph that shows a scene from a flower plantation. It is a marketing image that has the appearance of a random snapshot. The location is the site of production, and the flowers appear as they would in the plantation: large bunches of cut gypsophilia that have either been classified or are awaiting classification. It is notable that the company Flortec were responsible for bringing to Ecuador the variety 'super-gypso', which is well known for its very large buds. The meanings of advertisements are context-specific, and the significance of the inclusion of 'supergypso' in the image will only be understood by local growers.

The women who appear in the advertisement are neither there to represent abstract values as in Plate 7.1, nor to represent an invented image of what an Ecuadorian flower worker looks like, as in Plate 7.2. They appear in natural poses and are dressed as they would be in a flower plantation. In comparison to the model in the LoveRoses image (Plate 7.1), the limited makeup, the redness on the cheeks and the darker skin tone[5] would indicate to an Ecuadorian that these women are from rural areas. There is nothing to suggest that these women are not actual Ecuadorian flower workers caught on camera while classifying gypsophilia.

The 'snapshot' nature of the image could be attributed to a limited budget or lack of time. However, as the members of the company are internationally experienced businessmen, this is unlikely. The image must be considered in conjunction with the other elements of the advertisement. While the text conveys the message that this company offers technical expertise and international experience, the image imparts the message that this service is accessible and appropriate for flower growers in Ecuador. There is little depth in the photograph and therefore, we are given no indication of the size of plantation in which the women are working. As a result, no producer will reject the services offered as irrelevant to his/her business on the basis of hectares. The uniforms worn by the women in the picture suggest a certain level of professionalism, but this is tempered by the seemingly, disordered bunches of gypsophilia lying on the bench and the natural appearance of the workers in the image. The image offers a suggestion of greater professionalism through the services offered by the company, a message that is also represented in the text, while portraying familiar elements of small scale flower production in Ecuador in order to make the service on offer seem relevant to the target consumer.

Conclusions

The analysis of the advertising images presented here has shown the way in which representations of bodies are used within the creation of 'geographical lores' of origin in a variety of ways. There is an ever-shifting relationship between the depiction of landscapes and people in the visual creation of geographical knowledges for export commodities. In order to create an identity for the product, cut flowers are set in a particular imaginary landscape that aims to establish Ecuador as a site of production and therefore product identity. Each advertisement infuses the product with certain qualities by evoking an Ecuadorian identity through the juxtaposition of images of landscape, people and product. The role of the human image within this triangular commodity signifier is different in each advertisement. In the first advertisement the model serves to enhance the message conveyed by the landscape, and the meaning of the advertisement would remain stable even if the image of the model were removed. In contrast, the only element of the commodity-signifier indicating origin in the second advertisement is the image of woman and child – the portrayals of landscape and product are generic. In the third advertisement, there is an equilibrium between landscape, product and people as no element of the image could be removed without undermining the message conveyed regarding the location of production and the identity of the product.

The images of people in the three advertising images also differ in terms of what and whom they represent. In the first advertisement, the woman presented is an embodiment of a set of abstract qualities associated with flowers in general and more specifically with the version of 'Ecuadorianess' suggested by the landscape within which the product is positioned. The model is suspended between the Ecuadorian landscape and the product and serves to focus and emphasize the 'lores' of origin suggested by the wider setting. Like the Outspan girls (Mather and Rowcroft, Chapter 8, this volume) the model infuses the product with both general qualities (in the case of Outspan, 'fun' and 'health', and in the case of Love Roses, 'beauty' and 'fragility') and with an identity of source or nationality. The second advertisement similarly establishes a relationship between the landscape, the product and the two people within the frame of the advertisement. Rather than an embodiment of abstract values, the woman and child in this advertisement represent the social relations of production. Here it is the landscape that represents abstract qualities – in this case, 'nature' and 'fecundity'. The image of the product is similarly figurative rather than realistic. An intended focus upon the social development benefits of flower production results in the representation of worker and her family being the dominant image in the picture, and the Ecuadorian origins of the product are inscribed on the bodies rather than the product or the landscape. In the third image, the landscape, the product and the women shown in the advertising image all have equal value in so far as they interact to represent a more 'realistic' portrayal of the relationship between workplace, worker and product. Thus in addition to their multiple placements within a triangular relationship of signification, images of people play different roles within the creation of geographical knowledges of production. The three advertisements offer examples of bodies as embodiments of abstract values and bodies as sites inscribed with both mythical and realistic narratives of production.

In all three advertising images each body is marked by gender and ethnicity. These representations of gender and ethnicity serve to enhance both the qualities associated with the flower and the product identity conveyed by images of Ecuador. In both the first and the second advertisement, the juxtaposition of the female body and the cut flower recreates the imagined relationship between femininity and nature that has been well-documented elsewhere (Mies and Shiva, 1993; Warren, 1997). However, this relationship between femininity and nature is one that is intersected by ethnicity. The two advertisements link different versions of Ecuadorian femininity to different representations of nature, and thus recreate the linkages found in elite discourses of ethnicity and national identity in Ecuador. While a *mestizo* femininity is linked to a positive, nation-building fertility, an indigenous femininity is linked to an uncontrolled, destructive fecundity (Radcliffe and Westwood, 1996). In the advertising images it is the *mestizo* female who holds the arranged, processed bouquet of flowers, while it is the representation of an indigenous femininity in the second advertisement which is cast among a green, indistinct overgrowth. The neat, collage representation of the flower, however, and its link to child development suggest that flower work may curb an imagined indigenous fertility under the terms of community development. In the third advertisement, the image of rural female flower workers recreates the relationship with the product as found in the plantation, and thus supports the message of the applicability of Flortech's services to the smaller Ecuadorian grower that is conveyed by the advertising image as a whole.

Notes

1 Information regarding the marketing strategy for Ecuadorian flowers derives from an interview with an Expoflores representative in November 1996.
2 A *mestizo* identity results from the miscegenation of Spanish conquistadors and indigenous Americans at the time of colonization.
3 Colloredo-Mansfield (1998, 185) speaks of an 'urban, white–*mestizo* fear of filth and disease' which sees the indigenous body as the site where these qualities are located.
4 In a similar vein, a hotel in Quito created a stereotypical 'Indian' identity for its staff by taking elements of a particular rural community's identity and then changing those elements to appeal to the perceived tastes of North American tourists (Crain, 1996).
5 While some may balk at the suggestion of racial hierarchies of 'darkness' inferred by this phrase, the flower women and flower professionals I interviewed discussed ethnic identities as being linked to gradations of white and black. Women in the plantations associated an urban, *mestizo* femininity with whiter, smoother skin, and darker, sunburnt, wind chapped skin with rural women engaged in subsistence agriculture (Coulson, 2000).

References

Almeida, J. (1996) Fundamentos del Racismo Ecuatoriano, *Ecuador Debate*, 38, pp. 55–71.
Becerril, O. (1995) Como las trabajadoras agrícolas de la flor en Mexico hacen feminino el proceso de trabajo en el que participan? In Lara, S. (ed.), *Jornaleras, Temporeras y Boias-frias; El Rostro Feminino del Mercado de Trabajo Rural en America Latina*. Caracas: UNRISD, pp. 181–92.
Bordo, S. (1993) *Unbearable Weight: Feminism, Western Culture and the Body*. Berkeley: University of California Press.
Brandth, B. and Haugen, M. S. (2000) From lumberjack to business manager: masculinity in the Norwegian forestry press. *Journal of Rural Studies*, 16(3), pp. 343–55.
C.F.N. (Corporación Financiera Nacional) (1995) *Estudio del Sector Floricultor en el Ecuador*. Quito: Corporacion Financiera Nacional.
Colitt, R. (1996) Flower power. *U.S./Latin Trade*, February, p. 28.
Colloredo-Mansfield, R. (1998) Dirty Indians, radical indigenas and the political economy of social difference in modern Ecuador. *Bulletin of Latin American Research*, 17(2), pp. 185–205.
Cook, I., Crang, P. and Thorpe, M. (2000) Regions to be cheerful: culinary authenticity and its geographies. In Cook, I., Naylor, S., Ryan, J. and Grouch, D. (eds), *Cultural Turns/Geographical Turns*. New York: Prentice-Hall, pp. 109–39.
Coulson, J. (2000) *Embodying development: a study of female flower workers in Ecuador*, Unpublished PhD thesis, Department of Geography, University of Newcastle.
Crain, M. (1996) The gendering of ethnicity in the Ecuadorian Andes; native women's self-fashioning in the urban marketplace. In Melhuus, M. and Stølen, K. (eds), *Machos, Mistresses, Madonnas; Contesting the Power of Latin American Gender Imagery*. London: Verso, pp. 134–58.
Crang, P. (1996) Displacement, consumption, and identity. *Environment and Planning A*, 28, pp. 47–67.

De Chernatony, L. and McDonald, M. (1997) *Creating Powerful Brands: The Strategic Route to Success in Consumer, Industrial and Service Markets*. London: Butterworth-Heinmann.

De Oliver, M. (1997) Democratizing consumerism? Coalescing constructions of subjugation in the consumer landscape. *Gender, Place and Culture*, 4(2), pp. 211–33.

Foster, R. J. (1999) The commercial construction of 'New Nations'. *Journal of Material Culture*, 4(3), pp. 263–82.

Goldman, R. (1992) *Reading Ads Socially*. London: Routledge.

Haines, B. (1994) Ecuador: a high land of floral production. *FloraCulture International*, May/June, pp. 26–31.

Hendrickson, C. (1996) Selling Guatemala: Maya export products in US mail order catalogues. In Howes, D. (ed.), *Cross-Cultural Consumption: Global Markets, Local Realities*. London: Routledge, pp. 106–21.

Jackson, P. and Taylor, J. (1996) Geography and the cultural politics of advertising. *Progress in Human Geography*, 20(3), pp. 356–71.

Klein, N. (2000) *No Logo*. London: Flamingo.

Korovkin, T. (1998) Commodity production and ethnic culture, Otovalo, northern Ecuador. *Economic Development and Cultural Change*, 47, pp. 125–54.

Leslie, D. A. (1993) Femininity, post-Fordism and the 'New Traditionalism'. *Environment and Planning D, Society and Space*, 11, pp. 689–708.

Maharaj, N. and Dorren, G. (1995) *The Game of the Rose; The Third World in the Global Flower Trade*. Utrecht: International Books.

Meier, V. (1999) Cut-flower production in Colombia – a major development success story for women? *Environment and Planning A*, 31, pp. 273–89.

Mena, N. (1999) *Impacto de la Floricultura en Los Campesinos de Cayambe*. Cayambe, Ecuador: Instituto de Ecología y Desarrollo de las Comunidades Andinas (IEDECA).

Messaris, P. (1997) *Visual Persuasion: The Role of Images in Advertising*. London: Sage.

Mies, M. and Shiva, V. (1993) *Ecofeminism*. London: Zed Books.

Nayak, A. (1997) Frozen bodies: disclosing whiteness in Häagan-Dazs advertising. *Body and Society*, 3(3), pp. 51–71.

O'Barr, W. M. (1994) *Culture and the Ad: Explaining Otherness in the World of Advertising*. Oxford: Westview Press.

Palán, Z. and Palán, C. (1999) *Employment and Working Conditions in the Ecuadorian Flower Industry*. Geneva: Sectoral Activities Programme, ILO.

Radcliffe, S. (1999) Reimagining the nation: community, difference, and national identities among indigenous and mestizo provincials in Ecuador. *Environment and Planning A*, 31, pp. 37–52.

Radcliffe, S. and Westwood, S. (1996) *Remaking the Nation: Place, Identity and Politics in Latin America*. London: Routledge.

Rose, G. (1993) *Feminism and Geography: The Limits of Geographical Knowledge*. Minneapolis: University of Minnesota Press.

Stutzman, R. (1981) El mestizaje; an all inclusive ideology of exclusion. In Whitten, N. (ed.), *Cultural Transformation and Ethnicity in Modern Ecuador*. Urbana: University of Illinois Press, pp. 45–94.

van Liemt, G. (1999) The world cut flower industry: trends and prospects. Geneva: Sectoral Activities Programme ILO.

Warren, K. J. (ed.) (1997) *Ecofeminism: Women, Culture and Nature*. Indiana: University of Indiana Press.

World Bank (1995) Ecuador – agricultural census and information technology technical assistance project. Private Sector development, Project Id ECPA40106.

8 Citrus, apartheid and the struggle to (re)present Outspan oranges

Charles Mather and Petrina Rowcroft

There was a time not so long ago when it was impossible to eat a politically correct orange. Spanish oranges were handpicked by Franco, American fruit by brutalised Mexicans or Tom Joad in person, and nothing was more out than Outspan.

(*The Guardian*, 11 March 2000)

Introduction

This excerpt from *The Guardian* is the opening paragraph of an article describing a 'politically incorrect' skiing trip to Austria, following the election victory of the country's right wing Freedom Party. The significance of the paragraph for this chapter is not what it says, but what it omits about oranges and political correctness. While the author explains the problems associated with eating Spanish and American oranges – Franco and brutalized Mexicans respectively – the author does not explain why 'nothing was more out than Outspan'. We also know the origin of only two of the offending oranges, Mexico and the United States. The source of the third orange and its 'political incorrectness' is assumed to be known to readers of *The Guardian*. What is it about Outspan that allows the author to assume a common geographical knowledge about this system of provision?

Outspan was the brand name of all South African citrus exported by the country's Citrus Exchange between 1940 and 1998. It remains the brand for most of South Africa's oranges, grapefruits and lemons exported overseas. *The Guardian* piece reveals a key component of the commodity's lore: during the 1970s and 1980s Outspan was the target of the anti-apartheid movement, starting in the Netherlands and then spreading throughout Europe. As the brand for South African citrus, it was an obvious target for demonstrations and mobilization against apartheid. To this end the campaign against Outspan was used to spread knowledge about apartheid and its horrific impact on the majority of South Africans. Indeed, European consumers were given detailed knowledge of this system of provision and how its production and distribution depended on an institutionalized system of racial oppression. There is, however, more to Outspan's lore than its association with apartheid. When the anti-apartheid representations of Outspan are juxtaposed to the Citrus Exchange's own representations – both prior to and after apartheid – a complex struggle over the meaning and representation of the commodity is revealed. During the late 1960s and 1970s, Outspan commodity was promoted using a remarkable campaign that involved South African 'Outspan girls' (*sic*) and motorized vehicles resembling oranges. The lack of geographical knowledge associated with this campaign and the special training received by these women suggests that the Citrus Exchange was acutely aware of the potential danger of revealing the origins of Outspan oranges. The exporter's fears were realized in the early 1970s as the anti-apartheid movement appropriated the brand and aspects of the promotional campaign to unveil the social relations and production conditions of this system of provision. In the post-apartheid era the company responsible for citrus exports – Outspan International – has attempted to re-present the brand to European consumers. Its most recent promotional campaign, unveiled in 1995, suggests that it continues to be dogged by Outspan's lore. The campaign relies on a fascinating geographical imaginary, the meaning of which confirms that this object is firmly entangled in 'a web of wider social relations and meanings' (Jackson, 1999, 101).

The geographical knowledges used to promote the consumption of different commodities has been an ongoing focus of inquiry (Cook and Crang, 1996a; 1996b; Fine and Leopold, 1993; Leslie and Reimer, 1999). One of the central issues in this research is the 'spatial fetishism' associated with advertising, which is the source of Harvey's (1990) oft-quoted call to 'lift the veil' on the social relations and production conditions of the commodities we consume. Research inspired by the circuits of culture framework has extended the range of geographical knowledges associated with consumption; it has also stressed how this knowledge is constantly shaped and reshaped by a range of actors within the circuit. Crang (1996) has identified four kinds of constructed geographical knowledges associated with commodity circuits: those associated with 'sweeping geographical images' about a commodity; knowledges that reveal something of commodity production and distribution; knowledges about a commodity that derive from discursive associations; and finally knowledges about the appropriate setting for consumption. Advertisers use these geographical knowledges to 're-enchant' food and other commodities to encourage purchase and consumption. Crang (1996, 56) stresses that these geographical knowledges are not static, but instead are subject to constant reworking and reformulation through the participation of the 'full range of actors involved in the production, circulation and consumption of these meanings'. The reworking of geographical knowledges is possible, in part, because consumers do not rely solely on advertising imagery and promotional pamphlets for their knowledge of spatially distanced systems of provision. As Crang (1996, 54) writes, 'knowledges about commodities are not the preserve of promotional industries, but are constructed through rather more complex discursive fields'.

There are several innovative studies that reveal the geographical knowledges associated with advertising. Hendrickson's (1996) study of Guatemalan exports to the United States shows how advertisers represent and sell 'Guatemala' by drawing on ideas of tradition, authenticity and uniqueness. Clothing and other commodities presented in mail order catalogues are made by 'descendants of the Maya' in distanced places 'where life is thick with history and rich with tradition' (ibid., 113). Norton's (1996) study of tourist brochures of East African game reserves and Goss' (1993) analysis of similar material for Hawaii reveal how tour operators use a range of spatial imaginaries, which not only construct places but also indicate how they are to be consumed. While promotional material for Hawaii relies on a series of oppositions to modern North America, East Africa is represented as natural, primordial and even pre-historic (also see Cloke and Perkins, 1998). Crang's (1996) analysis of Boots' global bathroom products and Cook's (1994) reading of the geographical knowledges associated with tropical fruits are other examples of the maps of meaning that have come to be associated with contemporary consumption practices (see also Coulson, Chapter 7, this volume).

These case studies provide interesting insights into the different ways advertisers use geographical knowledges to promote or 'spatially fetishize' commodities. There are, however, very few studies that focus on how knowledges change or are challenged by different actors in the circuit.[1] Most of the work cited above presents a static analysis of how a set of promotional material is used to 're-enchant' a commodity ranging from a piece of clothing to an island in the Pacific Ocean. Moreover, very few accounts reveal how consumers might draw on a broader 'discursive field' to become informed about a system of provision. In other words, despite Crang's (1996) assertion that knowledges about commodities are constructed through broader discursive fields, case studies restrict themselves to analysing the representations produced by advertisers. The significance of the example discussed in this chapter is that the geographical knowledges associated with Outspan emerge from representations by the exporter and by a group of other actors located outside this system of provision. We argue that as these very different representations come into contact, they shape each other in a complex struggle over meaning and representation.

The first section of the chapter considers the origins of the Outspan label, several early local and international advertising campaigns and the use of the commodity by anti-apartheid campaigners. In the second section we examine Outspan's campaign after apartheid and explore its attempt to re-present citrus to European consumers.

Outspan

South African citrus growers have exported citrus to Europe since the early 1900s. However during the first two decades of the twentieth century the quality of the fruit was uneven. For industry experts, the source of the problem was an absence of coordination and quality control. Fruit transported from the interior of the country lacked adequate post-harvest treatment and was packed in ways that failed to prevent damage en route. At the ports, growers with smaller consignments found it difficult to secure shipping space and fruit was often stranded at ports for days and even weeks. A lack of infrastructure and an absence of coordination were identified as key reasons for the variable quality of fruit arriving in Britain, which was damaging the reputation of South African fruit more generally. In 1926 growers established the South African Cooperative Citrus Exchange (SACCE) to address problems of coordination and quality. The Citrus Exchange also was tasked with improving the marketing and promotion of citrus overseas and to this end it established an overseas office in London in 1930. By the end of the 1930s, the Exchange was exporting almost 80 per cent of South Africa's citrus crop, primarily to the United Kingdom.

The Citrus Exchange's promotional efforts initially were supported by the Empire Marketing Board, an organization established by the British Government to promote the consumption of goods produced in the Empire. Using newspaper advertisements and posters at retail agencies, consumers were urged to buy South African citrus as a sign of loyalty to the Empire. The slogans were unambiguous: 'Buy South African oranges and support the British Empire' (Cartwright, 1976, 6). The Empire Marketing Board's campaign lasted between 1926 and the onset of the depression in 1933. From the 1930s the Citrus Exchange also advertised South African citrus through its London office, but on a more limited scale and with a smaller budget. One of the problems it faced was that the fruit was packed and labelled using a wide range of different brand names and images, many of which evoked regional citrus producing areas in the country. Even though most of the brands included the 'Union of South Africa' as part of the label, the Citrus Exchange was finding it difficult to establish a single identity for South African citrus. The organization's difficulties in promoting South African fruit were compared to California's Sunkist brand, which was being successfully promoted in Britain. In 1937, the Citrus Exchange decided to follow the Californian growers' example and acquired the Outspan brand from an Eastern Cape citrus farm. A year later, all of the Exchange's citrus exported overseas was branded with the Outspan label and the use of regional or local brand names for fruit exported by the Citrus Exchange was no longer permitted.

While as much as 80 per cent of citrus exported to Britain carried the Outspan label, the Citrus Exchange remained a voluntary cooperative and it remained possible to export fruit outside of its marketing infrastructure using an independent brand name. The Citrus Exchange complained that fruit marketed outside of its infrastructure was of poor quality and was damaging the reputation of South African fruit. It was estimated that growers were losing as much as £100,000 a year as a result of poor quality fruit marketed by independent agents (Neumark, 1938). In the late 1930s, the Citrus Exchange was convinced that the solution to the problem lay in a single channel system for citrus exports from South Africa. In 1939 the Exchange persuaded the state to establish a Citrus Board and immediately nominated the Citrus Exchange as the country's only 'overseas distribution and marketing agent'.[2] From 1940, South African growers legally were compelled to export through the Citrus Exchange's single desk and for almost sixty years all of South Africa's citrus was labelled and marketed using the Outspan brand.

The Citrus Exchange did not take advantage of marketing the single brand for South African fruit until the 1960s. Overproduction of citrus – the result of a massive planting effort during the 1950s – was having a serious impact on the profitability of South African growers. Although the amount of citrus exported increased to almost 10 million cases, the price per case of fruit declined from R4.61 per case in 1958 to R3.25 in 1960. Despite the massive increase in volume the return to growers was the lowest in several years. The Citrus Exchange responded by hiring two sets of management consultants, one based in South Africa and the other in London. The consultants recommended a variety of changes to the structure and organization of the commodity chain. More significantly for the purposes of this study, they indicated the need for a 'dynamic selling organization' that would promote Outspan domestically and in Europe. An amount of R800,000 per year was budgeted for the overseas component of the campaign. By 1972 Outspan was spending more than R2 million a year promoting its fruit (Cartwright, 1976).

The domestic campaign was based on the distribution of Outspan peelers, juicers and other citrus related paraphernalia to white households in South Africa. Outspan dealers visited schools (also mainly white) in an effort to convince children of the health-giving attributes of oranges and other citrus products. In the mid-1960s the Citrus Exchange realized that the 'black market' remained 'untapped' and there were sporadic efforts to encourage black consumers to eat more oranges (cf. Rogerson, 1994). During the late 1960s the cooperative also marketed Outspan abroad using slogans ranging from 'the small ones are more juicy' to 'oranges: the summer fruit'. An Outspan hot air balloon was flown over London on a regular basis; a frequent landing spot during the summer was the city's Covent Garden Market to coincide with the delivery or sale of South African citrus. Outspan advertising was also posted on public transport buses as a way of promoting the health-giving and 'slimming' attributes of citrus. All of these efforts were overshadowed by the remarkable 'Outspan girls' (*sic*) campaign of the late 1960s and early 1970s, which involved young, white South African women touring various European countries during the northern hemisphere summer dressed in short skirt uniforms and wearing gaudy orange wigs. Their mode of travel around Europe was a small, motorized vehicle based on a Mini Morris chassis that resembled an orange. An entire fleet of 'Outspan-oranges-on-wheels' was constructed for the campaign and several working models still exist in Europe.[3]

Prior to a summer tour, the Outspan 'girls' were coached in the art of 'catching the public's eye' and selling the positive attributes of oranges and grapefruits. They also skipped 'about among the crowds – particularly crowds of children – distributing oranges and "fun" hats that bore the name "Outspan", and talked on what the Vitamin C in citrus fruit could do for the health of the average man and woman' (Cartwright, 1976, 96). When not at schools convincing children to eat oranges, these 'ambassadresses' of South African fruit distributed Outspan balloons to sick children in hospital or posed with well-known personalities like Stanley Holloway, star of the long running musical *My Fair Lady*. The tour itinerary was intensive: in Britain alone, these women visited 19 towns and participated in 'slimming galas' at each place. They also toured several other European countries including Switzerland, Germany, Italy and France. While the impact of the campaign on consumption is difficult to assess, from the point of view of the Citrus Exchange it had an outstanding impact on the profile of the Outspan brand. According to the author of a Citrus Exchange commissioned book, they made 'an immense contribution by their impact on the consumers who saw these bronzed and healthy girls wherever they went and associated them with the health-giving qualities of the fruit' (Cartwright, 1976, 97). An in-depth study of this campaign promises to reveal whether these 'bronzed and healthy Outspan girls' were able to convince 'pale-skinned' British consumers to purchase Outspan fruit (Mackenzie, 2003).

One of the significant aspects of the overseas campaigns of the late 1960s and 1970s was the absence of geographical knowledge about Outspan. Indeed, missing from all of the campaign material was the word 'South Africa': it did not appear on any of the promotional material produced by the Citrus Exchange's marketing experts nor on the jackets of the 'Outspan girls' which were dominated instead by oranges and the brand name. Moreover, there were no geographical images that might have indicated to consumers that this was South African fruit. Instead, the campaign focused on 'brand building', a process that involved identifying the product with health and vitality. Outspan oranges were an important source of Vitamin C and as a non-fattening snack, they were a way to become as thin as one of the 'Outspan girls'. The absence of any geographical knowledge or imaginary for Outspan in the campaigns of the 1960s and 1970s marked an important break with previous marketing efforts. It was noted earlier that the campaign organized by the Empire Marketing Board in the 1920s and 1930s referred directly to South Africa and its membership of the British Empire. As we shall see, the Citrus Exchange's decision not to use geographical knowledge to market its fruit was almost certainly a response to growing international condemnation of apartheid.

While the promotional campaign of the 1960s and 1970s may not have used spatial knowledges to sell Outspan, the product's lore was nonetheless steeped in South Africa's geography. This was, after all, the label for all South African citrus that had been exported to Europe since 1940. The word 'Outspan' is also distinctly South African and refers to the process of unyoking or unharnessing a team of oxen used to pull trek wagons. The term is associated with the movement of Afrikaner *voortrekkers* into the interior of the country during the 1800s and was regarded by the descendants of this minority as a heroic and defining moment of South Africa's history as a nation. Finally, although the 'Outspan girls' campaign did not indicate the origin of Outspan fruit, these women *were* South African and considered to be representatives not only of citrus but of South Africa too: 'South Africa owes a debt of gratitude to the Citrus Exchange for sending these young ambassadresses on tours abroad. They have sold not only our oranges but also the good qualities of South Africans and the good looks of our girls' (*sic*) (Cartwright, 1976, 96). That the Outspan girls represented a small minority of South Africa's population clearly escaped Cartwright, but not the anti-apartheid campaign against South African citrus.

While Outspan's own promotional campaigns might not have explicitly revealed the origins of Outspan or the conditions under which citrus was produced, the opposite was true of the anti-apartheid campaign of the 1970s and 1980s. The anti-apartheid movement's goal in Europe was to unveil the fetishism of this particular commodity and to 'tell the full story of [its] social reproduction' (Harvey,

1990, 423). As a commodity produced in South Africa, it was an ideal vehicle to promote the boycott of South African products. The anti-apartheid movement also used it to inform consumers about the violent impact of apartheid on the majority of the country's citizens. In the United Kingdom the anti-apartheid campaign against Outspan took various forms including pickets at supermarkets carrying this brand of fruit. In Ireland, supermarket employees refused to check out grapefruit originating from South Africa as part of the Irish Distributive and Allied Trades Union (IDATU) motion which called for the boycott of 'all South African goods and services' (*New Statesman*, 19 July 1985). Boycotts against Outspan throughout Europe coincided, significantly, with a sharp decline in the promotional campaign by the Citrus Exchange. By the mid-1980s the company had resorted to exporting fruit to Europe without Outspan labels, a move that reflected the success of this particular anti-apartheid campaign. Significantly, anti-apartheid action associated with Outspan wrested the representation of the commodity away from the company itself; indeed, consumers knowledge about the fruit identified in surveys by Outspan itself and by independent researchers probably has more to do with the efforts of anti-apartheid activists than the company's own efforts prior to or – as we shall see – after the demise of apartheid.

The origins of the European campaign against Outspan must be traced to the activities during the 1970s of the Dutch-based organization 'Boycot Outspan Aktie' (BOA, Boycott Outspan Action Foundation). BOA was one of many non-governmental organizations affiliated to the Anti-Apartheid Movement in the Netherlands. Although most of the groups associated with the movement focused on encouraging disinvestments from South Africa by Dutch-based corporations like Philips and Shell, the founders of BOA decided to boycott a very visible South African product: 'Because Outspan is so concrete and tangible, it can admirably be used to expose Western collaboration with apartheid' (Du Plessis, 1974, 61). Its efforts to encourage consumers to boycott Outspan relied on revealing the conditions under which citrus was produced in South Africa. But activists went beyond lifting the veil on the social and production relations of this particular commodity: they used the association between Outspan and South Africa as part of a broader campaign to inform consumers about the system of apartheid. To this end BOA produced a series of sophisticated publications, a monthly newsletter that was only phased out in 1994, a set of striking imagery to make consumers aware of apartheid and BOA-organized conferences to discuss the situation in South Africa based on the Outspan campaign. Publications ranged from shorter 'question and answer' booklets about the Outspan boycott (BOA, 1976) and the system of apartheid to detailed monographs complete with a list of additional readings (BOA, 1972). The title of one widely distributed monograph was *Outspan: Bouwstenen Voor Apartheid* (Outspan: Building Bricks for Apartheid) and is based on an a statement by BJ Vorster – president of South Africa in the early 1970s – who said 'Every time a South African product is bought, it is another brick in the wall of our continued existence' (BOA, 1972; Du Plessis, 1974, 59). This 80 page document – also published in English as *Outspan: Fruits of Shame* – begins by linking colonialism and the more formal system of apartheid to citrus production in quite specific ways: the arrival of settlers in the 1600s and the first citrus plantings, the expansion of citrus into the interior and the subsequent dispossession of African households, the forced removal of millions of Africans and the establishment of 'ethnic homelands', and detailed descriptions of the labour system that ensured white citrus farmers had access to cheap black labour (BOA, 1972).

BOA's first effort to boycott Outspan was in 1973 and given that South African fruit arrived in the Netherlands in June, their campaign against the product started some months earlier. It opened with a conference, organized in March 1973 in Leiden and the aim was to generate as much publicity about Outspan and apartheid as possible. The organizers of the conference attracted high profile anti-apartheid activists to speak at the conference including Ruth First, Peter Hain and James Phillips, a South African who had been banned because of his efforts with the South African Congress of Trade Unions. Panel discussions and strategies for the Outspan campaign followed presentations by these speakers. During the early months of 1973 BOA also contacted several large Dutch retailers and demanded that they discontinue selling Outspan fruit. The largest of these, Albert Heijn's, was the target of a national campaign during August 1973. Albert Heijn himself received a copy of the BOA bulletin informing him of the boycott and numerous posters were produced with 'ARE YOU ALSO AGAINST APARTHEID? THEN DON'T DO YOUR SHOPPING TODAY AT ALBERT HEIJN' and 'USE YOUR FREEDOM AS A CONSUMER! DO YOUR SHOPPING ELSE-WHERE TODAY. AGAINST APARTHEID? THEN TODAY NOT TO ALBERT HEIJN!!' (cited in Du Plessis, 1974, 66). A week before the campaign started Albert Heijn relented and for 'commercial reasons' announced that the supermarket chain would no longer be carrying Outspan oranges. Consumers in Holland were also made aware of the geographical origins of Outspan in an event organized in the town square of Veenendal, a town in central Holland. The town's 'Third World Store' mounted a map of South Africa and for 25 cents consumers could throw a rotten Outspan orange at the map; in return they were given a more 'wholesome' orange from another country. Four television crews, one from Britain, recorded this remarkable demonstration.

Perhaps the most interesting aspect of BOA's 1973 campaign was its response to the 'Outspan girls' campaign. BOA discovered that these young, white South African women were provided with a training that went beyond the health-giving and slimming attributes of citrus: they were also instructed not to discuss politics or the political situation in South Africa. Their interaction with consumers was to be limited to the fruit and what it could do for the health of consumers. BOA responded by organizing a group of 'multi-racial Inspan girls' who were to serve as a 'contrast to the exclusively white Outspan girls' (Du Plessis, 1974, 60). Unlike the 'Outspan girls' the 'Inspan girls' were instructed to engage in political discussion and were required to study South Africa and the system of apartheid. BOA's plan was to discover where and when the 'Outspan girls' would be performing and dispatch the 'Inspan girls' to provide consumers with detailed geographical knowledge about this system of provision. Sadly, perhaps, the

'Outspan girls' did not meet their BOA counterparts: at the beginning of July 1973 the marketing agents of the Citrus Exchange decided to skip Holland as part of their marketing tour of Europe. BOA's campaign in 1973 marked the end of Outspan's promotional campaign in Holland.

The imagery associated with the Dutch and other European campaigns was striking. One of the most widely used images shows a white hand squeezing a decapitated African head on a juicer (Plate 8.1). The white hand squeezing the African head is clearly wearing a business suit. The captions used with the image varied from 'Outspan bloedsinappels' (Outspan blood oranges) to 'Pers geen Zuid-Afrikaan uit', which translates literally as 'Don't squeeze a South African dry'. Mounted on placards, the image was used to publicize conferences and to discourage shoppers from purchasing oranges at retail outlets. Members of BOA also used the images at the main ports and wholesale markets where Outspan fruit was being offloaded or sold. A second image associated with the Outspan campaign has a black man imprisoned behind the bars of an Outspan orange raising his fist in the 'Amandla' (power) salute, an important signifier of the struggle against apartheid (Plate 8.2). The phrase 'Nee tegen Outspan sinaasappelen uit Zuid-Africa' means 'No to Oustpan oranges from South Africa'. These images provided startling visual representations of citrus and apartheid that complemented the detailed information provided to consumers on this system of provision.

BOA's campaign in the Netherlands had an important impact on the Citrus Exchange's ability to export to this country. Most supermarket chains refused to stock Outspan for fear of boycotts. Demonstrations at wholesale citrus markets using the 'Outspan blood oranges' images in cities like Rotterdam provided additional complications to the Citrus Exchange's efforts to enter this market. Assessing the impact of boycotts against Outspan in Europe, which spread throughout the continent during the 1970s and 1980s is, however, more complicated. Trade figures do suggest that the organization found it difficult to operate in an environment where its brand was the target of a very active campaign. Through the 1970s and 1980s average citrus exports from South Africa were between 3.2 and 3.9 million tons of fresh fruit. In the 1995/6 season, several years after the withdrawal of sanctions, citrus exports had almost doubled to 7.5 million tons (FAO, 1999). Although South African growers had lost ground to Australia and Argentina during the 1980s, in the 1995/6 season it exported half of all citrus exported from the southern hemisphere. The rapid rise in exports after apartheid suggests that while the boycotts did not stop citrus from being consumed, it did have an impact in limiting the expansion of South Africa's citrus industry. The Outspan campaign also would have played a role in contributing to the growing number of European countries imposing a ban on South African agricultural products. In 1985, for instance, Sweden passed a special law prohibiting the importation of agricultural products from South Africa. A year later Ireland passed a similar law barring imports of South African food and fibre commodities. The Commonwealth Heads of Government, with the exception of the United Kingdom, agreed in 1986 to ban imports of South African food products. While the United Kingdom remained open to Outspan

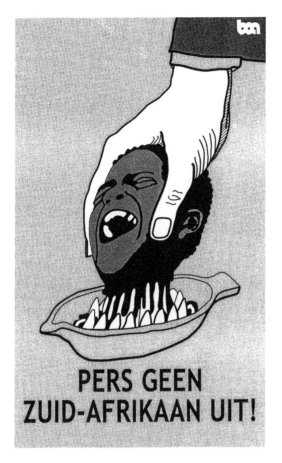

Plate 8.1 Don't squeeze a South African dry.
Source: BOA, 1972.

branded oranges, the anti-apartheid campaign was successful enough to force the Citrus Exchange to sell fruit without any identification.

The significance of the campaign from a 'circuits of culture' perspective is that the BOA campaign wrested the representation of this commodity out of the hands of the Citrus Exchange. From a commodity associated with young white women and Vitamin C, it was reworked for consumers as a product that originated in apartheid South Africa. Indeed, as BOA was at pains to point out, the production of South African citrus depended on a system that was based on dispossession and a brutal labour regime. The focus of the campaign was to reveal to consumers the geographical knowledge associated with this system of provision: from the pamphlets and monographs of citrus and apartheid to the demonstrations involving a map indicating the origins of Outspan fruit, the campaign's goal was to *locate* Outspan in apartheid South Africa. The boycott of

Plate 8.2 No to Outspan oranges from South Africa.

Source: BOA, 1976.

Note
Boycott Outspan Action Foundation – Organization now defunct and original artist cannot be traced.

Outspan appears to have played an important role in consumer's geographical knowledge about the product. Citrus Exchange surveys (Coetzee, 1998) and a recent independent survey conducted in Kent suggest that a surprisingly high percentage of people associate Outspan citrus with South Africa, even though they might not know much about past or contemporary production relations on citrus farms. Poole and Baron's (1996) survey of 300 citrus consumers in Kent revealed that as many as 60 per cent of consumers knew that they had purchased citrus from South Africa. This level of consumer knowledge about origins in citrus was second only to Spain, where the 'recall' was 70 per cent. Consumer knowledge about Outspan revealed in surveys like these is probably the result of a very active campaign organized by civil society rather than the promotional efforts of the Citrus Exchange.

Re(presenting) South African citrus

The campaign against Outspan, as noted in the previous section, marked the rapid downscaling of various marketing campaigns in Europe by the Citrus Exchange. By the end of the 1980s, its marketing efforts had all but ceased. From an active effort aimed at 'brand building' in the 1960s and 1970s, the Citrus Exchange was now selling fruit without labels. In the early 1990s, the organization underwent a transformation in line with the broader liberalization of the South African economy. The Citrus Exchange changed its name to Outspan International and it transformed itself from a cooperative into a private company. The purpose of these changes was to create an organization that could regain ground lost through the sanctions and boycotts of the 1970s and 1980s. A key element of this strategy involved 're-defining' the Outspan brand through a new advertising campaign.

Outspan decided that the new campaign launched in 1995 would be designed and coordinated in South Africa. Apart from the major initiatives of the 1970s, most advertisements for Outspan had been designed by European based agencies for European consumers. The logic was that what might appeal to a German consumer would be different from what might convince a French shopper to purchase an Outspan orange. The organization's decision to launch the campaign from South Africa may have been an attempt to emulate successful marketing efforts by Australian and New Zealand beer and wine companies during the 1990s (Merrett and Whitwell, 1994). What is unusual about the Outspan campaign was its decision to use a single set of images for all of its export markets including Europe, North America and Asia (Jackson, 1999). Executives at Outspan felt that the brand was comparable to other globally recognized names such as Nike, Coca-Cola and McDonald's; as one of the 300 most important brands in the world, it deserved a set of images that corresponded to its brand profile and its ambition as a global supplier of fresh citrus (Coetzee, 1998).

The images of the 1995 campaign focused on the theme of wild African animals with oranges, lemons, grapefruit and other citrus cultivars acting as the sun, a watering hole or a flower. Although the campaign is the same for all of the organization's overseas markets, there is some flexibility in the production of the images. The advertisements were distributed to European, North American and Asian representatives on a CD-Rom. The software programme on the CD-Rom allowed the overseas agent to select from several different landscape scenes, several different animal species, a range of citrus cultivars and a few key slogans. For example, a male lion relaxing under an acacia could be replaced with a kudu or an elephant; similarly a grapefruit or a Valencia orange could replace a lemon representing the sun. With the exception of the general slogan appearing at the bottom of the images 'Outspan: the source of refreshment', slogans were interchangeable but only with a standard set of phrases held on the CD-Rom. For instance 'grapefruit the shape fruit' could be replaced with the more general slogan of 'citrus the health fruit'. The programme was not, however, flexible enough to allow the agents to change the key geographical imaginary of the campaign.

A striking aspect of the images is the absence of both people and citrus trees. The social relations of production associated with a particular commodity are rarely revealed in promotional material (Fine and Leopold, 1993). When these relations are revealed, as in the case of Guatemalan clothing produced for North American consumers, they are reconstructed and encoded with messages of traditionalism, authenticity or uniqueness (Hendrickson, 1996). In the case of South African citrus it is difficult to imagine how pictures of poorly paid workers housed in shacks would promote the consumption of citrus fruit. Moreover, it is difficult to see how images of the highly industrialized nature of citrus production – intensive use of water, the application of chemicals for fertility, pest and weed control, and the mechanized packing and intensive post-harvest treatment of citrus – would convince consumers that it is in their interests to eat Outspan fruit. In several important ways the images convey precisely the opposite message: this is fruit that is juxtaposed to 'nature' (cf. Crang, 1996); indeed, it is literally *located in nature* in an epistemology that separates society and nature into distinct conceptual fields. It is fruit that appears not to have required human labour or technology in its production.

Several devices are employed to ensure that the fruit is part of a 'natural system'. First, the fruit is placed in a 'natural' setting where people are absent and where the production relations of citrus remains obscure. The fruit is located in African game reserves, places of sanctuary from society; these are spaces where nature has been reconstructed and externalized in its 'original' and unchanging form, where animals can exist in an environment unimpeded by the influence of humans (cf. Katz and Kirby, 1991). A second device to establish the 'naturalness' of the fruit is the use of oranges, grapefruit and lemons – commodities that owe their origin to industrial farming practices – to portray natural features of the environment including the sun, a watering hole, fruit in a tree or a flower on a water lily. By replacing 'natural' features with citrus, the fruit itself takes on natural attributes. The imminent consumption of 'natural features' by animals – a giraffe drinking water/grapefruit and an elephant eating a flower/tangerine – represents a third device that attests to the fruit's natural qualities. By consuming the fruit in these forms, the animals testify to its naturalness.

The encoded message of nature depends of course on the landscapes within which the fruit has been located. As noted earlier, these landscapes represent Africa's game reserves: wide open plains with lions, kudu, giraffe and rhinoceros, punctuated with acacia trees, are place myths associated with the better-known East African parks of Serengeti, Ngorogoro Crater and Amboseli. They are the images – minus the fruit of course – which European consumers see on 'nature' documentaries produced by David Attenborough (Burgess, 1990) or in tourist brochures of East Africa (Norton, 1996). The framed images also may correspond to a personal experience or memory of an East African safari, although Norton's (1996) survey suggests that the imaginary may not always correspond to the tourist experience (also see Urry, 1995). Significantly for our concern with geographical origins, the image of savannah plains is a key Western geographical imaginary for the continent of Africa as a whole; these images have become the 'sights and sounds we instinctively associate with wild Africa' (Adams, 1992, 42). Indeed, as Adams (1992, xiii) has pointed out 'Ask someone to paint a picture of Africa today and it would resemble…the Serengeti…thousands of wildebeest marching nose to tail in a line a mile long, while several well-fed lions laze under a flat topped acacia tree nearby.' The use of African imaginaries associated with travel and tourism points to the enduring links between the tourist gaze and consumption practices or as Crang (1996, 52) writes, the 'touristic quality of much contemporary consumption'. The safari is the African experience par excellence (Pieterse, 1992) and Outspan's efforts to promote South African citrus represent an attempt to 're-enchant' this particular commodity by tying it to a real or anticipated tourist experience (Lash and Urry, 1996). Several of the phrases in the images – especially 'peel the exotic taste of summer' – combined with cloudless skies and obviously hot temperatures promise an eating experience that is not only natural, as argued above, but also allows consumers to associate a travel experience with their own domestic culinary regimes (Cook and Crang, 1996a).

When interpreted as a series of texts, the Outspan images are strikingly similar to a set of written texts about Africa. In an article on post-apartheid travelogues, Crush (2000) argues that South African writing about the rest of the continent constructs two Africas, one that is urban, decaying, violent and filthy. The other Africa reflects the countryside and descriptions of this rural landscape are favourable and rely on the themes of emptiness, silence and an absence of boundaries; the landscapes are deeply aestheticized and are described with deep hues and colours. Crush (2000, 447) draws on Pratt's (1992) *Imperial Eyes* and the travelogues themselves to suggest that the 'Edenic vision' in these narratives is 'deeply rooted in colonial discourses and ideas about (African) nature and landscape'. If the travelogues explored by Crush represent South Africa's written fiction about Africa, the Outspan campaign would appear to represent one aspect of this fiction in visual form. Like post-apartheid writing of the countryside, the Outspan images are un-peopled and represent a stark contrast to the wars, famines and droughts that dominate CNN images of Africa. The images lack boundaries in that the vistas are unbroken by fences, barriers or anything else that is human made. Outspan's landscapes are also highly aestheticized and visually striking with clear blue skies and the dark greens and browns of the savannah. That the Outspan images are for the consumption of northern audiences suggests that this post-apartheid binary vision of 'an essentialised African continent that varies little from country to country, place to place' (Crush, 2000, 447) is also one that may correspond to consumers outside of post-apartheid South Africa (Pieterse, 1992).

It is significant that Outspan did not choose to use a South African geographical imaginary or place myth for Outspan fruit. Instead, the brand is associated with a set of African imaginaries that may include South Africa, but which do not distinguish this country as the place of origin of the Outspan brand. In interviews with Outspan executives it was difficult to uncover precisely why they had not selected a place or person imaginary that would have clearly indicated that the fruit was South African. When we suggested that they could have used Cape Town's Table Mountain or Nelson Mandela eating an orange on Robben Island (cf. Lash and Urry, 1996) their responses were evasive and unconvincing. One of the managers involved in the campaign claimed implausibly that the slogan 'The source of refreshment' referred to South Africa's democratic transition. A far more plausible argument is that Outspan's decision to use a set of pan-African imaginaries relates to the commodity's turbulent lore. Rather than attempting the difficult and potentially dangerous task of engaging with this lore, Outspan has displaced the geographical knowledge associated with the product commodity elsewhere. Indeed, it is likely that the campaign of the late 1960s and early 1970s as well as this more recent campaign were shaped by a discursive regime that existed around oranges and apartheid. In the earlier campaign, the absence of geographical knowledge was almost certainly related to the growing rejection of apartheid and anything associated with this system of racial oppression. In the late 1990s, promotional efforts continued to be shaped by the events of the 1970s and 1980s when the anti-apartheid movement successfully appropriated the label as a way of revealing the social relations associated with the production of South African citrus. Rather than using a geographical imaginary associated with South Africa, it used a much safer African imaginary. Outspan's campaigns have always struggled to create meaning for the product in the face of extremely powerful and convincing alternative geographical knowledges.

Conclusion

This chapter has explored the geographical knowledges associated with Outspan oranges. Significantly, the most powerful representations of this spatially distanced system of provision appear to have been produced not by the owner of the brand, but by the European anti-apartheid movement. Their realist representations focused on Outspan oranges as the product of apartheid South Africa and the volume of detailed information provided about this system of provision is astounding. From in-depth analyses of the land and labour situation in South Africa, to detailed knowledge on wage rates and gender divisions of labour on citrus farms, the movement unveiled the anatomy of this commodity chain. The exporter's representations prior to and after apartheid were shaped by these powerful images and detailed geographies. Their attempts to depoliticize the 'Outspan girls' and the absence of geographical imaginaries in the campaign of the 1960s and 1970s reflected a concern over growing international rejection of apartheid. When the South African exporter was prepared to use a geographical imaginary to represent their fruit after apartheid, we argue that this too was shaped by Outspan's lore. Rather than attempting to re-present South African citrus, and thus engage with earlier representations of the fruit, the commodity was linked to a far safer pan-African set of geographical imaginaries.

The struggle to rework the meaning of this commodity to European consumers, and the prominent role played by actors outside this system of provision, confirms Crang's (1996) argument about the broader discursive regime from which consumers may draw knowledge about commodities. In the case of Outspan these representations engage with each other in fascinating ways, illustrated most vividly in the Dutch anti-apartheid movement's black 'Inspan girls' a stark opposite to the exporter's own white Outspan 'girls'. And while the exporter's campaign of the 1970s failed to convey the origins of Outspan citrus, the anti-apartheid campaign revealed precisely where and under what conditions fruit was produced. The discursive regime around Outspan between the late 1960s and mid-1990s was characterized by considerable struggle and contestation over representation and meaning.

Notes

1 A study that does focus on the changing lore of a product, although not its *geographical* lore, is Hebdige's (1988) study of the Vespa motor scooter.
2 It is not unusual for growers in a country to coordinate fruit exports, although the level of state regulation differs between fruit producing countries. For a comparison of marketing arrangements for southern hemisphere apple producers see Roche *et al.* (1999); for more detail on Outspan and its structure prior to 'deregulation' in 1998 see Mather (1999).
3 For a view of what these cars looked like see: http://www.westhouse.demon. co.uk/orange.htm

References

Adams, S. (1992) *The Myth of Wild Africa*. London: Unwin.
BOA (Boycot Outspan Aktie) (1972) *Outspan: Bouwstenen Voor Apartheid* (Outspan: building bricks for apartheid). Nijmegen: Stichting Studentenpers.
BOA (Boycot Outspan Aktie) (1976) *Vragen en Antwoorden over de Boycot Outspan Aktie*. Nijmegen: Stichting Studentenpers.
Burgess, J. (1990) The production and consumption of environmental meanings in the mass media: a research agenda for the 1990s. *Transactions of the Institute of British Geographers*, 15, pp. 139–61.
Cartwright, A. P. (1976) *Outspan Golden Harvest: A History of the South African Citrus Industry*. Cape Town: Purnell.
Cloke, P. and Perkins, H. C. (1998) Cracking the canyon with the awesome foursome: representations of adventure tourism in New Zealand. *Environment and Planning D: Society and Space*, 16, pp. 185–218.
Coetzee, J. (1998) Building a super brand. Unpublished paper presented at the 1998 Extension Conference, Sun City, South Africa.
Cook, I. (1994) New fruits and vanity: symbolic production in the global food economy. In Bonanno, A., Busch, L., Friedland, W. H., Gouveia, L. and Mingione, E. (eds), *From Columbus to Conagra: The Globalisation of Agriculture and Food*. Lawrence: University of Kansas.
Cook, I. and Crang, P. (1996a) The world on a plate: culinary culture, displacement and geographical knowledges. *Journal of Material Culture*, 1, pp. 131–54.
Cook, I. and Crang, P. (1996b) Commodity systems, documentary filmmaking and new geographies of food: Amos Gitai's Ananas. Paper presented at the 1996 Annual Conference of the Institute of British Geographers/ Royal Geographical Society, Glasgow, January 1996.
Crang, P. (1996) Displacement, consumption and identity. *Environment and Planning A*, 28, pp. 47–67.
Crush, J. C. (2000) Africa unbound: redemption and representation in the new South African travelogue. *Tijdschrift voor Economische en Sociale Geografie*, 91, pp. 437–50.
Du Plessis, E. (1974) Don't squeeze a South African dry. *Africa Today*, 21, pp. 59–68.

FAO (Food and Agriculture Organization) (1999) Citrus fruit fresh and processed: annual statistics, 1999. CCP:CI/ST/99, Food and Agriculture Organization of the United Nations, Geneva.

Fine, B. and Leopold, E. (1993) *The World of Consumption*. London: Routledge.

Goss, J. D. (1993) Placing the market and marketing place: tourist advertising of the Hawaiian Islands. 1972–92. *Environment and Planning D: Society and Space*, 11, pp. 663–88.

Harvey, D. (1990) Between space and time: reflections on the geographical imagination. *Annals of the Association of American Geographers*, 80, pp. 418–34.

Hebdige, D. (1988) *Hiding in the Light: On Images and Things*. London: Routledge.

Hendrickson, C. (1996) Selling Guatemala: Maya export products in US mail-order catalogues. In Howes, D. (ed.), *Cross-Cultural Consumption: Global Markets, Local Realities*. London: Routledge.

Jackson, P. (1999) Commodity cultures: the traffic in things. *Transactions of the Institute of British Geographers*, 24, pp. 95–108.

Katz, C. and Kirby, A. (1991) In the nature of things: the environment and everyday life. *Transactions of the Institute of British Geographers*, 16, pp. 259–71.

Lash, S. and Urry, J. (1996) *Economies of Signs and Space*. London: Sage.

Leslie, D. and Reimer, S. (1999) Spatializing commodity chains. *Progress in Human Geography*, 23, pp. 401–20.

Mackenzie, C. (2003) *The 'Outspan girls'*. MA thesis, University of the Witwatersrand, Johannesburg.

Mather, C. (1999) Agro-commodity chains, power and territory: re-regulating the South African citrus export filière. *Geoforum*, pp. 61–70.

Merrett, D. and Whitwell, G. (1994) The empire strikes back: marketing Australian beer and wine in the United Kingdom. In Jones, G. and Morgan, N. (eds), *Adding Value*. London: Routledge.

Neumark, S. D. (1938) *The Citrus Industry of South Africa*. Johannesburg: Witwatersrand University Press.

Norton, A. (1996) Experiencing nature: the reproduction of environmental discourse through safari tourism in East Africa. *Geoforum*, 27, pp. 355–73.

Pieterse, J. N. (1992) *White on Black: Images of Africa and Blacks in Western Popular Culture*. New Haven: Yale.

Poole, N. D. and Baron, L. (1996) Consumer awareness of citrus attributes. In Trienekens, J. H. and Zuurbier, P. J. P. (eds), *Proceedings, 2nd International Conference on Chain Management in Agri- and Food Business*. Wageningen: Department of Management Studies, Wageningen Agricultural University.

Pratt, M. L. (1992) *Imperial Eyes: Travel Writing and Transculturation*. London: Routledge.

Roche, M., McKenna, M. and Le Heron, R. (1999) Making fruitful comparisons: southern hemisphere producers and the global apple industry. *Tijdschrift voor Economische en Sociale Geographie*, 90, pp. 410–26.

Rogerson, C. M. (1994) Unnecessarily draining the consumer's pocket: an historical geography of advertising in South Africa's black urban townships. *South African Geographical Journal*, 76, pp. 27–32.

Urry, J. (1995) *Consuming Places*. London: Routledge.

9 Tropics of consumption
'Getting with the fetish' of 'exotic' fruit?

Ian Cook, Philip Crang and Mark Thorpe

Introduction

Our chapter begins with two different perspectives on tropical, or 'exotic' fruit. A first example is taken from the 1967 Walt Disney cartoon *The Jungle Book*, an adaptation of Rudyard Kipling's 'Mowgli' stories. In the film, the song 'Bare necessities' was performed by the 'long established TV, radio, night-club and film star Phil Harris' whose mannerisms 'dominated the picture's indolent character of Baloo, the bear, the jungle's easy-going layabout who is a past master at the art of avoiding work and effort' (Anon, 1967). In singing Bare necessities, Baloo attempts to pass on this 'art' to the 'man-cub' Mowgli: why toil for your necessities when, here at least, 'Mother nature' has done the work for you? Paw paws and prickly pears can simply be picked from her trees, ripe and ready to eat.[1]

A second example derives from the poem *A Kumquat for John Keats* by Tony Harrison (1981). Imagining a meeting with Keats, Harrison plans how he might persuade the long-dead poet to rethink the intense (but unproblematic) pleasure Keats claimed to derive from tasting, judging, nibbling, scrunching, sucking and carving various fruits in a letter to his beloved Fanny (Dietrich, 1991). As Harrison put it:

> For however many kumquats that I eat
> I'm not sure if it's the flesh or rind that's sweet,
> and being a man of doubt at life's mid-way
> I'd offer Keats some kumquats and I'd say:
> You'll find one part's sweet and one part's tart:
> say when the sweetness or sourness starts.

The performance of 'Bare necessities' draws upon a Golden Age scenario in which residents of the colonial tropics were able to live a life free from toil in a place where the fruit literally fell from the trees. This particular part of the film recently was used by the UK supermarket chain Tesco in marketing papayas and mangoes. Pictures of Baloo appeared on shelf markers that proclaimed that 'Bare Necessities' could be bought for, perhaps, 99 pence each. In contrast, Harrison's poem uses the contradictory and confusing sensations experienced in eating a kumquat to characterize the pleasures and pains of (his) life, the difficulty in identifying any clear dividing line between one and the other, and the highly ambivalent feelings which accompany any act of consumption (ibid.).

This chapter examines issues surrounding commodity fetishism raised by these two 'fruity' examples. It is about the ways in which businesses have attempted to market tropical fruits to UK consumers since the mid 1980s. It is about the ways in which this has involved the reworking of colonial discourses regarding tropical lands and their inhabitants. It is about the often ugly histories which can emerge in the process, and how these histories can be fleshed out and (re)attached to the fruits. It is an attempt to 'get with the fetish' (Taussig, 1992) of 'exotic' fruits, to 'hijack' them, to re-contextualize them, and to illustrate how this process can contribute to wider political/academic debates about commodities and their geographies. We contend that commodity fetishes are by no means neatly woven 'veils' which simply mask the origins of consumer goods. Rather, these veils are prone to *mundane* rupture and recombination in the everyday lives of consumers and business personnel (e.g. Cook *et al.*, 1998b, 2000; Gillespie, 1995; Rand, 1995) and to *strategic* rupture and recombination in the work of NGOs, activists, educationalists and other culture workers (e.g. Christian Aid, 1996; Cook and Crang, 1996; Cox, 1992; Hartwick, 1998, 2000; Klein, 2000; Mather and Rowcroft, Chapter 8, this volume).

This is a *strategic* essay, an attempt to contextualize, rupture and re-contextualize 'exotic fruitiness': first, in the UK advertising and fruit (trade) press since the mid-1980s; second, within justifications for European tropical colonialism from the sixteenth to nineteenth centuries; and finally in discussions of Carmen Miranda's 'fruity' star persona from the 1940s to date. In so doing, we seek to contribute to a wider political/academic project whose aim is to show that, if we pick up a paw paw, a prickly pear, a kumquat, any commodity, we have in our hands a bundle of social, cultural, political, economic, biological, technological, geographical, historical and other relations which ensured that it travelled from those places, to that shelf, in that form, at that time and at that price (Bonanno *et al.*, 1994; Goodman and Watts, 1997; Watts, 1999). Our prime purpose is to use this approach to unpack and re-work representations of 'ethnic', 'exotic' places, people and products, and to develop hybrid, multicultural geographical imaginations which might be (re-)attached to commodities; imaginations which might add extra dimensions to their consumption.

Tropical fruit in the trade press

Since the 1970s, the rapid growth in the global sourcing of foodstuffs and cooking knowledge has meant that fresh and processed fruits and vegetables from all over the world have become commonplace in British supermarkets. Produce constructed as 'exotic' by mainstream retailers now appears alongside 'out-of-season' produce which is also grown thousands of miles away (Anon, 1990; Friedland, 1994; Goodman and Watts, 1997; Mackintosh, 1977). In the UK fruit trade press, developments in fresh produce sections of supermarkets were constructed as fundamental to the changing shape of food retailing in the late 1980s and 1990s. In most supermarkets it is the fresh produce section that is encountered first by shoppers (Willis, 1991). This is the only section of the store where 'consumers not only tolerate, but encourage, each other to handle the food' (Litwack, 1989, 43; MSI, 1988) and where product lines are stocked simply 'in order to say that they are there' (Litwack, 1989, 46). 'The area has become a showcase for many retailers, helping to form consumer impressions about the[ir] overall merchandising effort' (Fengler, 1989, 48) and to create an image of the retailer being in touch with the latest 'glamorous fashions' in food retailing (Anon, 1986a; Elman, 1988). Given their temperature-controlled perpetual springs, the cornucopia of fruits and vegetables readily available on their shelves, and no sign of any work having gone into their production, it is not unusual to find these fresh produce departments described as 'consumer paradises' (e.g. Gilbert, 1988, 47).

The fruit trade press has made much of the ways in which multiple retailers have transformed low volume speciality fruits such as kiwis, pineapples and avocados into everyday shopping items. It has suggested that a similar process could be repeated for other low volume fruits such as mangoes and papayas (Anon, 1987; Henderson, 1992; Moore and Utterback, 1996). As one British produce buyer suggested, 'the greatest antidote to fear is knowledge and it is extremely important to present as much information as possible to the consumer' (in Anon, 1986b). Retail chains now provide a range of point-of-sale materials, including free information and recipes printed on stickers attached to individual fruits, in-store live or video demonstrations and book(let)s on sale in both supermarkets and in mainstream bookstores (Anon, 1986a,b; Cook, 1994; Elman, 1988; Gilbert, 1988). Not only do these materials provide instructions on how to ripen, slice and introduce tropical fruits into the culinary worlds where they will end their lives, but they also place them within imaginary 'exotic' worlds from which they supposedly originate.

In the books read and used by supermarket buyers, these latter worlds are of 'natural abundance' in which fruits come from highly sanitized 'origins' (Cook, 1994, 1995). Historical stories are usually set out in opening paragraphs describing each fruit. These begin with a brief discussion of Latin names, and then present an account of the invariably European male naturalists who 'discovered' the fruit during his travels in the tropics, named it after his patron or himself, and brought samples back to Europe to be grown in hothouses. One book, for example, notes that the Fortunella species of kumquat 'was introduced into Europe (from China, Japan and Malaya) in 1846 by Robert Fortune, who spent in all 19 years in the Far East collecting plants for the Royal Horticultural Society. This little genus...was named in his honour' (Johns and Stevenson, 1985, 149). A more expansive story is often told if a fruit's 'discovery' and/or displacement can be linked to more famous explorers and scientists like Christopher Columbus or Joseph Banks. Columbus plays a part in the story of lemons, sugar cane, pineapples and papaya (Anon, n.d.a; Johns and Stevenson, 1985; Market News Service, 1989; Touissant-Samat, 1992) while Banks is said, for example, to have 'suggested that Captain Bligh should be sent to Tahiti in the Bounty to collect breadfruit seedlings and take them to the West Indies, where it was hoped, they would provide a staple diet for the Negro slaves' (Johns and Stevenson, 1985, 77). Still other fruits are traced to their biblical and missionary 'origins': 'there is a legend that [the banana] was the forbidden fruit and that [its] leaves were the first garments of the first man and his mate' (ibid., 57).

Such 'origin' stories are by no means only historical. Connections also are made between particular fruits and touristic imaginations of 'bounty' most often associated with the contemporary tropics. Tropical fruits are often represented as 'coming to you' like Baloo's 'bare necessities'. In advertising and packaging materials, they are usually falling out of the trees (e.g. Plate 9.1), available in baskets placed beneath them (e.g. Plate 9.2) and/or presented to the viewer by

Plate 9.1 Photograph of Libby's *UmBongo* carton, manufactured by Nestlé.
Source: Nestlé.

Plate 9.2 Photograph of a publicity folder used by 'Geest Tropical' staff.
Source: Geest plc.

indigenous women (see following section). The most common icon in these representations is the palm tree, swaying whole over the fallen fruits or providing the leaves upon which prepared fruits are laid out. Related written accounts often present fruit as part of simple, unified and absolutely different 'tropical cultures' which are in close harmony with their natural surroundings (e.g. where 'the *fertile plains* yield large crops of banana, coconut and sugar cane' (Anon, n.d.b)). Here, it is 'nature' that produces these fruits, not the labour (and other inputs) through which nature is transformed in agricultural production. When agricultural production is mentioned, tropical culture/nature discourses are often used to justify the intervention of multinational capital. One trade press story, for example, overturns the *West Side Story* portrayal of Puerto Rico as an 'ugly island [...] of tropical diseases' with the suggestion that the country now is 'a tropical paradise combining the warmth of Latin temperament with the efficiency of American know-how and modern technology' (Anon, 1988, 33). Here, it seems, the 'natural' state in the tropics – that dangerous paradise – can be best brought to fruition by 'rational' capitalist intervention.

Finally, the relations between the (imaginary) worlds of tropical production and temperate consumption also draw on the movements of tourists and migrants between them. The experience of consuming tropical fruits often is associated with travelling as a tourist to the tropics, mixing with locals and sampling their delicacies in 'authentic' settings (Friedman, 1991; Kale *et al.*, 1987). But, as outlined in one British newspaper, such an 'authentic' setting can be recreated at home after a trip to the shops:

> [You can] turn your kitchen into a Caribbean cookhouse and treat yourself to some tastes you've never tried before....Travel is the theme this spring, but if you can't get away to the fascinating places you've been reading about, you can at least cook up a little of the atmosphere in your own home, with recipes concocted from combinations you've never dreamed of, using unusual ingredients *plucked straight from the tropics*. There are so many of these weird and wonderful foods available from the shops and street markets, you have only to look for them. Then put some heat into spring with these exotic feasts.
>
> (Anon, n.d.c, 113, emphasis added)

Thus tourism to and migration from the tropics are linked: one fruit book introduced readers to 'the new-found wealth of curious and intriguing fresh produce daily arrayed before us in street markets, shops and supermarkets' by stating how 'Britain's multiracial communities have brought the flavours of the world into our high streets. ...Make the most of them – you need not go to India, Singapore or the West Indies – *stay at home and relish them here*' (Heal and Allsop, 1986, 1, emphasis added).

In advertising and other knowledges produced to encourage their purchase, then, paw paws, prickly pears, kumquats and other fruits are often inserted into a tropical commodity fetish. This fetish is inhabited by a particular combination of people and 'nature'. It includes the naturalists and explorers who 'discovered' many of these fruits in their travels to the tropics in the heyday of European imperialism and who, along with wealthy patrons, often gave their names to them. It also incorporates the 'people of the tropics' with whom 'British' consumers supposedly interact either on holiday or in Britain's 'multicultural communities'. Their 'culture' is the 'other' which these consumers can supposedly 'get a bit of' through buying these fruits – although the resulting 'commodification of otherness', it is argued, reinforces a divisive 'us'/'them', 'British'/'ethnic' mentality (Attar, 1985; hooks, 1992; May, 1996a,b). Finally, and perhaps most importantly, this fetish is bound up with the tropics themselves: the climate, landscape, rich soils and lush vegetation which yield these fruits, apparently without assistance, as they have done as far back as anyone can see. This fetish appears to obscure the fact that fruit is commercially farmed in the tropics by people, and often under circumstances which consumers would find disturbing (Cook, 1994, 1995; Cook *et al.*, 2000; Watts, 1993). However not everything is hidden. There are ruptures *within* this fetish – its stories of imperial adventure, notions of multiculture, and representations of tropical 'nature' – which can be further explored and rewoven to give these fruits some historical depth and thereby a less 'exotic' and more ambivalent meaning *on the surface*.

Tropical fruit in European colonial discourse

In this section, we take hold of particular ruptures in this tropical commodity fetish – such as mentions of Columbus, Banks, palm trees, paradise, fruit falling from the trees, fertile plains yielding crops, and temperament tamed by know-how – in order to flesh them out and recombine them in an alternative web of 'origins'. We concentrate upon the roles which fruits and their (lack of) cultivation in the tropics played in discourses of 'discovery' (re)produced by European explorers, botanists, colonists and philosophers from the fifteenth to nineteenth centuries, and consider how they helped to justify European colonial expansion. An appropriate starting point is Columbus's (1992 [1493]) widely circulated and influential *Letter* which described what he had 'discovered' in the West Indian tropics. This founding journey/text is said to mark 'the beginning of the modern era (as from then on the world (was) closed and mankind (had) discovered the totality of which it was part' (Mason, 1990, 6) and to have precipitated 'the most astonishing, most intense, and most genocidal discovery within the history of human exploration' (ibid.). Moreover, Columbus presented European philosophers with new knowledges they could not afford to ignore (ibid.; Hulme, 1986, 1990; Todorov, 1984), and helped to shape a colonial imagination whose discourses and practices are still with us, within and beyond the worlds of tropical fruits.

In the absence, both of linguistic and botanical understandings of this 'New World', Columbus' interpretation of the lands and people he 'discovered' was shaped by the cultural baggage he and his crew brought with them, including European discourses (in)directly gleaned from the Bible, Marco Polo's accounts of Far-Eastern trade, Herodotus's accounts of Greece's 'barbaric' neighbours, and other writings which were well known among the upper classes in late Medieval Europe (Hulme, 1986; Mason, 1990; Todorov, 1984). In the centuries that followed the publication of Columbus' *Letter*, countless other sources were read into what he supposedly 'discovered'. In all readings, however, this 'New World' was inhabited by two distinct types of people: darker-skinned, 'savage', cannibalistic 'Caribes' and lighter-skinned, simple and accommodating 'Arawaks'. The dubious grounds for this discrimination are well established: for example, those who accepted the European presence were called 'Arawaks' and those who resisted it were labelled 'Caribes' (see Hulme, 1986; Mason, 1990; Todorov, 1984). However our interest is in the subsequent representation and experience of the 'New World' tropics by colonizing Europeans as both 'heavenly' and 'hellish', and in the ways in which its fruits became central to justifications for many of their actions.

'Heavenly' accounts of the tropics were based on representations of the 'Arawak' people and of the natural landscapes with which they lived in apparent harmony. In Columbus' *Letter*, the riches of the 'West Indies' were represented as ripe for the taking as their 'natives' were either guileless and generous (the 'Arawaks') or containably monstrous (the 'Caribes') (Hulme, 1986). A standard feature in this and subsequent accounts was the bearing of fruit and other gifts of food into the colonizing gaze by Arawak women who 'were the first who came to give thanks to Heaven and bring whatever they had, especially things to eat, such as bread made from ajes, peanuts, and five or six kinds of fruit' (Columbus, 1987, 144). Accounts of generous hospitality were explained by 'natural abundance': the 'Indians' lived in a 'land...so rich that it is not necessary to work much to feed themselves or clothe themselves, so they all go naked' (ibid., 139). The wide circulation of these accounts meant that this radically different 'New World', with its 'lack of property relations...[and] absence of restrictive morality', acquired a 'firmly utopian flavour' in sixteenth century Europe (van den Abbeele, 1984, 45; Mason, 1990; Porter, 1990). Two centuries later, European naturalists continued to draw upon Columbus' accounts to contrast the 'steril[e] desolation' of the Old World and this New World whose 'Nature appears more active, more fruitful, we may even say more prodigal of human life' (Manthorne, 1984, 376).

During the nineteenth century, textual accounts of this utopian relationship between nature and human life became bound up with changing ways of representing landscapes in Europe. Rather than conforming to the ideals of Italianate landscape composition, artists like William Hodges sought more 'accuracy' in their work by 'selecting and arranging the parts to express [a landscape's] essential and unique qualities' (ibid., 378). The reading of biblical discourses into this combination of exploration, botanical science and landscape art led the palm tree to become a key element in these landscapes because:

> although not named specifically in Scripture, the Tree of Life became historically identified with the date palm indigenous to the biblical lands. But, as exploration of the American tropics proceeded from the fifteenth century onward, the coconut palm – which provided all the earthly needs of the native 'Indians' – began to appear more likely as the tree to which biblical accounts alluded.
>
> (Manthorne, 1984, 376)

Those wanting to represent the tropics as a 'paradise on earth', an 'Eden', therefore invariably painted a coconut palm into, and the people of the tropics out of, the picture. In this way European 'social fantasies [...] of harmony, industry, liberty, unalienated joie de vivre all [became] projected onto the non-human world' (Pratt, 1992, 125).

Although represented in scientific accounts and landscape painting as (almost) uninhabited, this 'New World' was obviously full of people, and colonial experiences often involved extremely violent struggles over land (las Casas [1552], 1992). This was the 'hellish' side of the coin. In the centuries following the publication of Columbus' *Letter*, questions about the rights to possess and live off this land formed the subject of heated political and philosophical debate in Europe. Most important, perhaps, was the debate over whether the people 'discovered' there (particularly the 'Caribes') could be considered 'human'. While liberal commentators such as Descartes considered 'Indians' to be 'uncivilised' people with whom Europeans could work, others such as Locke judged them through the ways in which they used their land (Seed, 1993). Summarizing this latter view, Hulme (1990, 30 and 33, emphasis added) argues:

> the central division...lies between those who 'improve' and those who merely 'collect': only the former are fully rational and therefore fully human.... Tahe earth produces fruits and feeds beasts. If the Indian merely collects what is provided 'spontaneously' then it is not easy to see how he differs from the beasts that are similarly provided for; except of course that as a human being he is guilty of ignoring the law of nature that is incumbent upon humans but not animals.... Certainly those who can be seen as 'having renounced Reason' by shedding blood, or even those who have revealed an 'enmity' towards a man of reason, may, under Locke's fearsome gloss on the *law of nature*, 'be treated as Beasts of Prey', those 'dangerous and noxious Creatures, destroyed for the same reason that he may kill a *Wolf* or a *Lyon*'.

Such argument about the (ir)rational use of land and indigenous people's reactions to its appropriation by European colonists became a major justification for conquest in the Americas from the sixteenth century onwards. Much of this was accomplished through encoding land as female, virginal and as something to be rescued from the 'savage natives' who resisted its colonization (Mason, 1990; Montrose, 1991). Even 'Arawak'-style hospitality was treated as a threat – it was suspected that food was being offered to fatten up colonists for the cannibal feast. Thus even friendliness could legitimately be met with violent 'self defence' (Hulme, 1986; Mason, 1990; Pieterse, 1992).

From 1492 onwards, these unfolding tropical encounters were always more than a harmless matter of their representation as Heaven and Hell on Earth. Through centuries of exploitation and resistance, the clashing of desires (religious, commercial, territorial, sexual), terror, confusion, certainty, bluffing, double-bluffing, disease and weaponry led to the mass enslavement, rape and murder of indigenous people throughout the New World in what Richard (1991, 59) has described as the 'greatest genocide in the history of humanity'.

The 'New World's' indigenous population of approximately 100 million people in 1492 was slashed to between ten and twelve million in just 15 years (ibid.). The region's history of extreme colonial violence and exploitation later included the millions of enslaved African people who died on marches to slaving ports, in the middle passage, and in their 'seasoning' on plantations, as well as others who survived the brutalities of plantation life (James, 1938; Patterson, 1969). Yet representations of the 'New World' tropics as a Golden Age Paradise persisted. After his tour of the West Indies in 1858, for example, Trollope described the typical recently emancipated inhabitant as:

> idle, unambitious as to worldly position, sensual and content with little.... He lies under a mango-tree, and eats the luscious fruit in the sun; he sends his black urchin up for breakfast and behold the family table is spread. He pierces a cocoa-nut and lo! there is his beverage. He lies on the grass surrounded by oranges, bananas and pine-apples.
>
> (in Pieterse 1992, 199; see also Anon, 1921; Springhall, 1970)

The discursive separation of tropical commodities from the people who cultivated them therefore appears to be as old as global capitalism itself. Centuries of colonial violence and exploitation have been both justified and hidden in constructions of the tropics as a place where fruit simply falls from the trees and nobody does any work. Golden Age scenarios continue to be reproduced in the marketing of tropical fruits. Past and present are connected in heavenly and hellish ways: the tropical commodity fetish is the tip of an historical-geographical iceberg.

Tropical fruit in the star persona of Carmen Miranda

Recontextualizing these commodities as the fruits of Heaven and Hell is important, but it is not enough. A complementary and more constructive recontextualization is required. In this section, our intention is to grasp some important ruptures in both the exoticized worlds of fruit marketing and in the colonial discourses outlined above – ethnic stereotypes in the marketplace, the easy appropriation of other cultures, tropical landscapes as female bodies, the yielding of fruit, heaven and hell – and to flesh these out in a further context which works through the ambivalent pleasures characterized in Harrison's *Kumquat* poem. We have chosen the 'star persona' of the Brazilian actress Carmen Miranda (Plate 9.3) because of the way in which her role in the marketing of United Fruit's Chiquita bananas in the USA (the company's 'Miss Chiquita' is based upon Miranda) has been discussed by Cynthia Enloe (1989). Enloe attempts to 'unveil' the commodity fetish of the Chiquita banana by tracing the sexual politics and political economies of its production in the Central American tropics and argues that Miranda's role was to contribute to an 'exotic' smokescreen that masked these ugly politics. Miranda's movies 'helped make Latin America safe for American banana companies' because she 'personified a culture full of charm,

Plate 9.3 Carmen Miranda.

Source: © Twentieth Century Fox/Kobla Collection.

unclouded by intense emotion or political ambivalence (Enloe, 1989, 124). Like the bananas she wore on her head, Miranda was exotic but mildly amusing' (ibid., 127) and she helped to present a 'harmoniously exploitative' relationship between the United States and Latin America in which the latter happily 'shared' its riches with the former (see also Burton, 1992; López, 1993; Roberts, 1993). However, closer readings of Miranda's film roles and their consumption throw a different light on this argument, one in which her 'fruity' persona allowed her to 'get in touch with the fetish' of North American representations of Latin American exoticism and to channel its energies in critical (perhaps 'revolutionary') directions.

The ways in which Hollywood studios represented the idealized relationship between North America and Latin America in the 1940s was shaped by the US government's 'Good Neighbour' policy and parallels more recent constructions of 'sharing' and 'acquisition' in the Euro-American tropical fruit trade. In films made for audiences North and South of the border, the extraction of wealth by companies based in the former at the expense of the latter was heavily downplayed, while stereotypes of the inseparable 'nature' of the tropics and its peoples, and a feminization of 'nature' as 'a willing target of appropriation' were centre stage (Burton, 1992, 38). The 'authentically exotic' Latin American woman became an essential film character, but complex cultural, political and economic compromises had to be made in choosing the right actresses:

> Hollywood (and the United States) needed to posit a complex 'otherness' as the flip side of wartime patriotism and nationalism and in order to assert and protect its economic interests. A special kind of 'other' was needed to reinforce the wartime national self, one that – unlike the German or Japanese 'self' – was non-threatening, potentially but not practically assimilable (that is, non-polluting to the purity of the race), friendly, fun-loving, and not deemed insulting to Latin American eyes and ears [...] What Hollywood's good neighbour regime demanded was the articulation of a different female star persona that could be readily identifiable as Latin American (with the sexual suggestiveness necessary to fit the prevailing stereotype) but whose sexuality was neither too attractive (to dispel the fear of attraction or miscegenation) nor so powerful as to demand its submission to a conquering American male.
>
> (López, 1993, 70–71 and 73)

Miranda was perhaps able to negotiate this role most successfully, with the help of film director Busby Berkeley, whose gigantic and 'garishly delirious' musicals provided an ideal vehicle for her 'fruity' star persona (Murray, 1971; Rubin, 1993). She took 'as her costume enormous flowers, fruits, and vegetables intermixed with exaggerated Brazilian dress, (and thereby became) the image of an overflowing cornucopia of South American products, ripe, ready, and eager for picking by North American consumers' (Roberts, 1993, 14). Miranda acted 'fruitily' in the sense that audiences often saw her as crazy, odd, full of rich quality, and suggestive. Her performance in Berkeley's 1943 film *The Gang's All Here* (the only Miranda film to which Enloe refers) is perhaps her most famous. Miranda steals the show in three spectacular, hallucinatory night club scenes, including a performance of 'The lady in the tutti frutti hat'. This sequence has been described as 'an orgy of giant phallic bananas and vulval strawberries through which Miranda moves with a controlled chaos' (Dibbell, 1991, 44; Roberts, 1993, 15). Nothing in cinema 'has captured more accurately the North Atlantic fantasy of the South as a site for innocent erotic anarchy' (Dibbell, 1991, 45).

Miranda's simultaneous reproduction of and resistance to colonial stereotypes of Latin American ethnicity and femininity has been the subject of much debate. Latin American ethnic stereotypes were reproduced through such performances, the gendering of this tropical landscape as female (and this female as a tropical landscape) was plain to see, and fruits were centre stage. Yet this chaotic world was one in which the controlling character was a Latin American woman performing stereotypical 'exotic' culture in such an exaggerated and comically knowing way that she/it became difficult to appropriate by a colonizing gaze. This was vividly illustrated in the ways in which contemporary North American film critics made sense of her performances. Some tried to fit her into discourses of tropical nature:

> in terms of the physical, of the body – as wild, savage, and primitive, like an exotic animal, 'enveloped in beads, swaying and wriggling macaw like...,
> skewering the audience with a merry mischievous eye'. *Motion Picture* compared her to 'a princess out of an Aztec frieze with a panther's grace, the plumage of a bird of paradise and the wiles of Eve and Lileth combined'. Beyond the usual savage and bestial qualities, the fanzine suggests that the fall of man might also be attributable to Miranda.
>
> (Roberts, 1993, 10)

However as a 1941 review indicates, this was no easy task:

> What it is that Carmen has is difficult to describe; so difficult, in fact that dramatic critics have grown neurotic in their attempts to get it into words that would make sense and at the same time not brand them as mad sex-fiends. Nevertheless, it must be attempted again. First, there is the impact of Carmen's costumes,...always covering her thoroughly with the exception of a space between the seventh rib and a point at about the waistline. This expanse is known as the Torrid Zone. It does not move, but gives off invisible emanations of Roentgen rays. (Her songs') words are absolutely unintelligible to a North American..., but what the listener hears – or hopes he is hearing – is unmistakable.
>
> (cited in Roberts, 1993, 11)

Rather than considering Miranda as 'exotic yet mildly amusing' (Enloe, 1989, 127), these reviews suggest that her Hollywood persona could be 'exotic' but profoundly disturbing. Reviewers were used to capturing artists' performances through their writing, but many found that Miranda's hyper-sexuality and hyper-ethnicity made her difficult to capture.

This, as many have argued, was her intention. Through displacing 'her florid femininity...onto her costume and make-up, the drag queen's tools of first resort' (Dibbell, 1991, 45), Miranda 'acknowledged and openly participated in her (own) fetishisation' but 'knowingly' stared back at the audience implicating them in the process of constructing ludicrous stereotypes of Latin American 'ethnicity' and 'femininity' (López, 1993, 76). Miranda has been seen to have not quite 'burst the illusory bubble of the Good Neighbour, but by inflating it beyond recognition she highlight(ed) its status as a discursive construct' (ibid., 78). For more marginal cinema audiences feeling similarly stereotyped in terms of their gender and sexuality, Miranda's performances 'offered resistant, alternative viewing options' and opened up a discursive space in which to express and confirm their own subjectivities (Roberts, 1993, 19). In 1940s America, straight female and gay male fans began to appropriate new combinations of cosmetics, hats, clothing and altered accents to 'perform the Miranda masquerade' (ibid.).

Miranda's kitsch value also was a transnational phenomenon, requiring the construction of a different and intersecting history (Konder, 1982). This history stretches well outside Hollywood's cinematic frame, and acknowledges the extroverted multiculturalism of her persona which was not so much within, but between, cultures both in its production and consumption. Before her rise to Hollywood fame, Miranda was a star of Brazilian film, radio and stage (Roberts, 1993). It was this 'authentic' Brazilian talent which Hollywood moguls desired, but without regard to what her act had meant to these Brazilian audiences. The costumes and hats which became her Hollywood trademarks were based on those worn by black women of the Northern Brazilian city of Bahia (Dibbell, 1991), which she had adopted and exaggerated for a wider Brazilian audience in her rise to national stardom in the 1930s.

In contemporary intellectual circles, Brazil was reconciling itself to a multiracial heritage which included 'the large, disempowered African-Brazilian population, which asserted its presence partially through the acceptance of the music and dance traditions of samba and candomblé' (Roberts, 1993, 12). Miranda's stage act therefore portrayed a powerful Afro-Brazilian femininity:

> The candomblé temples of Rio were established by women, freed slaves from Bahiam, the sixteenth-century capital of Brazil. These women, Bahians, were stereotyped within Brazil as women with shawls, turbans and flirtatious ways. In addition, 'Their special relationship with the old continent was...recognised.... They knew the region, they had 'samba in the foot', they had survived, and they kept the culture going'.
>
> (ibid., 13)

A white Brazilian woman adopting and exaggerating the garb of an Afro-Brazilian woman to construct a character on stage (or in carnival) was 'as Brazilian as rice and beans. Racial [and gender] cross dressing had become a model of authenticity' (Dibbell, 1991, 44; Roberts, 1993). In her dress and behaviour, Miranda was not naïvely reproducing discourses though which European colonists justified the invasion of Brazilian lands and lives. Although the Hollywood publicity machine attempted to represent Miranda as a 'naïve Latina, awed by North American culture' (Roberts, 1993, 8), strong traces of her former incarnation remained.

Finally, it is important to note how Miranda's fruity star persona more recently has been resurrected as a symbol of resistance in Brazilian popular culture. In the late 1960s, she/it was drawn into an emerging *Tropicalismo* movement that aimed to oppose a highly nationalist and repressive Brazilian government by promoting an extroverted, multicultural Brazilian identity which she had represented in the

1930s. *Tropicalismo* involved the development of a distinctive musical/performance culture whose critical edge had to be hidden in double and triple entendre to circumvent the censors. The movement:

> saw in itself an extension of ideas dating from the Brazilian avant garde of the '20s and '30s when Brazilian culture had been defined as anthropophagic, or cannibalistic, devouring European, African and Native American elements and expelling them in a truly new world culture.... (Its) insights were focussed on the clash between the archaic culture of rural Brazil and the mass produced culture of the cities. *Tropicalismo* provided a new perspective on local and international kitsch and made striking use of rhythms of Bahia and the other northeastern states that had been relegated to the status of folklore.
>
> (Lindsay, 1989, n.p.)

Tropicalismo attempted to 'articulate a broad movement of opposition to the military government' (Rowe and Schelling, 1991, 96), and to parody those 'myths which represented Brazil as a tropical paradise' (Roberts, 1993, 14). Given Miranda's legendary use of the double and triple entendre, her performances of Brazilian culture as hybrid and extroverted, the way she had drawn marginalized Bahian musical traditions into urban Brazilian culture, and her presence as icon of local and international kitsch, she became a key figure in the movement. '*Tropicalismo* appropriated her as one of its principal signs, capitalising on the discomfort that her name and the evocation of her gestures could create' (Veloso, cited in Roberts, 1993, 14), a discomfort which was crucial to a movement whose intention was to 'fuck with whatever hierarchies it could get its hands on' (Veloso, cited in Dibbell, 1991, 45).

Returning to our initial critique of the veils of commodity fetishism neatly concealing exploitation, it is important to emphasize how Miranda (alongside her many fans and imitators) magnified elements already present in the tropical commodity fetish she was supposed to inhabit. She was the subject of a colonizing gaze that made particular connections between nature, people and culture in the tropics. She inhabited a fetish where culture was nature, nature was female, and each willingly offered its fruits to the outside world. Yet her embodiment and knowing exaggerations of these stereotypes, her persistent use of double and triple meanings, her humour, and her sheer 'in your face' tropical 'fruitiness' meant that she was no passive subject of that colonizing gaze. The extent to which she 'got with this fetish' has been seen as something from which a great deal can be learned. As Dibbell (1991, 43 and 47) has argued:

> identity and difference, the central problems of (her) life, have become the central problems of our politics..., and they are problems we seem doomed to take at once far too seriously and not seriously enough. Perhaps camp, the refined art of being serious about the frivolous and frivolous about the serious, is just the finesse we need to tiptoe through the mine field of multiculturalism...and come out alive.

Conclusion

We began this chapter by promising to spin a number of arguments out of, and through, the tropical commodity fetish within which the arrival of 'exotic' fruits on British supermarket shelves has been framed. We argued that through analysing the role of fruit in European colonial discourse and Carmen Miranda's 'star persona', we could show how this commodity fetish could be ruptured and (some of) its elements recombined in ways which 'rub taken-for-granted history against the grain' (Simon, 1994, 131). Building on Taussig's (1992) assertion, we argued that 'getting with the fetish' in this way can redirect its energies in more critical – perhaps even 'revolutionary' – directions. We stated that this approach would lead us into wider debates about commodities and their geographies.

We would reiterate that the commodity fetish within which tropical fruits' 'origin' stories are often set share individual symbols, and an overall symbolic structure, with representations of and justifications for European colonial endeavours in the tropics. Ways of representing relations between people, nature and culture in the tropics – and the resultant 'advantages' of Western intervention in these relations – can be seen in both contexts. Thus this commodity fetish places centre stage centuries of colonial exploitation. Today's international trade in tropical fruits owes a great deal to this history. It is a history in which material and symbolic processes were so thoroughly interwoven that calls to 'get behind the veil, the fetishism of the market and the commodity, in order to tell the full story of social reproduction' (Harvey, 1990, 423) are calls to sweep aside processes which are fundamental to this story. Therefore if (elements of) these relationships are recognized as occupying centre stage, those committed to strategically rupturing commodity fetishes should not only try to replace this (false) geographical imagination with that (real) one. They/we also can usefully work with what's there, thickening, extending, reworking, 'getting with' material-semiotic connections which are visible and apparent.

We have argued elsewhere that the manufacture, marketing and consumption of 'ethnic' or 'exotic' foods simultaneously relies upon, reproduces, and breaks down dominant 'them' and 'us' distinctions between 'non-white' and 'white' Britons (Cook *et al.*, 1999, 2000). There seems to be plenty of scope – even a pressing need – for research into the geographies of commodities to include thoroughly entangled, seriously mischievous, strategically fruity research and writing which works through such cultural/economic ironies. Carmen Miranda the person, the star persona, and the international icon might be an inspiration here. Recontextualizing commodities can be an important strategy for critical geographers who want to flesh out stories of cultural and economic power (Fusco, 1995; Gómez-Peña, 1993; Nash, 2000; Walker *et al.*, 1994). These stories are sitting on supermarket shelves, fresh and ready to pick. With a bit of help, stories about tropical fruits can provide a vivid and fleshy way into understandings of our multicultural pasts, presents and their power-soaked entanglements. These fruits can be shown to embody multicultural/historical/economic geographies worth celebrating and commemorating. Like a kumquat, simultaneously sweet and tart, they provide 'pleasure' (with) an edge and a bite (Gaines, 1990, 7).

Acknowledgements

This paper comes from ESRC project (R000236404) 'Eating places: the provision and consumption of geographical food differentiations', so thanks go to them and to the many colleagues and audiences who have chipped into it over the years.

Note

1 See the Bare necessities' lyrics at http://www.fpx.de/fp/Disney/Lyrics/The Jungle Book.html (accessed 23/10/03).

References

Anon (n.d.a) Commodity report: Papaya. *Eurofruit*, 31, n.p.

Anon (n.d.b) *Postcard from Jamaica*. Kingston, JA: Novelty Trading Company.

Anon (n.d.c) Tropical treats. Pre–1992 British Newspaper Sunday Supplement.

Anon (1921) *The World and Its People: Little Folks of Other Lands*. London: Thomas Nelson and Sons.

Anon (1967) Sleeve notes for the LP record. *Songs from Walt Disney's The Jungle Book and Other Disney Animal Favourites*. London: EMI/Music for Pleasure.

Anon (1986a) A first for NIFP with exotics. *Fruit Trades Journal*, 8 August, p. 9.

Anon (1986b) Exotic produce sales are now booming. *The Grocer*, 18 January, p. 13.

Anon (1987) A formula for exotics. *Fruit Trades Journal*, 16 October, p. 8.

Anon (1988) Year-round Puerto Rican exotics offer promise in UK. *Fruit Trades Journal*, 6 May, p. 33.

Anon (1990) Fresh fruit and vegetables. *The Ethical Consumer*, April/May, pp. 10–19.

Attar, D. (1985) Filthy foreign food. *Camerawork*, 31, pp. 13–14.

Bonanno, A., Busch, L., Friedland, W. H., Gouveia, L. and Mingione, E. (eds) (1994) *From Columbus to ConAgra*. Lawrence: University Press of Kansas.

Burton, J. (1992) Don (Juanito) Duck and the imperial-patriarchal unconscious. In Parker, A., Russo, M., Sommer, D. and Yaeger, P. (eds), *Nationalisms and Sexualities*. London: Routledge.

Christian Aid (1996) Are you sure the fruit you buy is full of goodness? Campaign advertisement in *The Guardian*, 31 October, p. 13.

Columbus, C. (1987) *The Log of Christopher Columbus*. Southampton: Ashford Press.

Columbus, C. (1992) The letter of Columbus (1493). In Hulme, P. and Whitehead, N. (eds), *Wild Majesty*. Oxford: Clarendon Press.

Cook, I. (1994) New fruits and vanity: the role of symbolic production in the global food economy. In Bonanno, A., Busch, L., Friedland, W. H., Gouveia, L. and Mingione, E. (eds), *From Columbus to ConAgra*. Lawrence: University Press of Kansas.

Cook, I. (1995) Constructing the exotic: the case of tropical fruit. In John Allen and Doreen Massey (eds), *The Shape of the World*. Oxford: Oxford University Press, pp. 137–42.

Cook, I. and Crang, P. (1996) The world on a plate: culinary culture, displacement and geographical knowledges. *Journal of Material Culture*, 1, pp. 131–53.

Cook, I., Crang, P. and Thorpe, M. (1998a) Eating translations: food, victual knowledges and the incorporation of rupture 'eating places', Working Paper No.1, Department of Geography, UCL.

Cook, I., Crang, P. and Thorpe, M. (1998b) Biographies and geographies: consumer understandings of the origins of foods. *British Food Journal*, 100, pp. 162–7.

Cook, I., Crang, P. and Thorpe, M. (1999) Eating into Britishness: multicultural imaginaries and the identity politics of food. In Roseneil, S. and Seymour, J. (eds), *Practising Identities: Power and Resistance*. Basingstoke: Macmillan.

Cook, I., Crang, P. and Thorpe, M. (2000) Regions to be cheerful. Geographies of culinary authenticity. In Cook, I., Crouch, D., Naylor, S. and Ryan, J. (eds), *Cultural Turns/Geographical Turns*. Harlow: Longman.

Cook, I. *et al.* (2000) Social sculpture and connective aesthetics: Shelley Sacks' 'Exchange Values'. *Ecumene*, 7, pp. 338–44.

Cox, C. (1992) *Chocolate Unwrapped*. London: Women's Environmental Network.

Dibbell, J. (1991) Notes on Carmen: Carmen Miranda, seriously. *Village Voice*, 29 October, pp. 43–5.

Dietrich, C. (1991) The raw and the cooked: the role of fruit in modern poetry. *Mosaic*, 24, pp. 127–44.

Elman, D. (1988) Consumer expenditures study: produce. *Supermarket Business*, 43(9), pp. 169 and 218.

Enloe, C. (1989) Carmen Miranda on my mind: international politics of the banana. *Bananas, Beaches and Bases*. London: Pandora Press.

Fengler, B. (1989) Produce departments cry help. *Supermarket Business*, 44, pp. 1–11.

Friedland, W. (1994) The new globalisation: the case of fresh produce. In Bonanno, A., Busch, L., Friedland, W. H., Gouveia, L. and Mingione, E. (eds), *From Columbus to ConAgra*. Lawrence: University Press of Kansas.

Friedman, J. (1991) Further notes on the adventures of phallus in blunderland. In Nencel, L. and Pels, P. (eds), *Constructing Knowledge*. London: Sage.

Fusco, C. (1995) *English is Broken Here*. New York: New Press.

Gaines, J. (1990) Introduction: fabricating the female body. In Gaines, J. and Herzog, C. (eds), *Fabrications: Costume and the Female Body*. London: Routledge.

Gilbert, L. (1988) Exotic produce shouldn't be a mystery. *Supermarket Business*, 43, pp. 47–50.

Gillespie, M. (1995) *Television, Ethnicity and Cultural Change*. London: Routledge.

Gómez-Peña, G. (1993) *Warrior For Gringostroika*. Saint Paul, MN: Graywolf Press.

Goodman, D. and Watts, M. (eds) (1997) *Globalising Food*. London: Routledge.

Hartwick, E. (1998) Geographies of consumption: a commodity chain approach. *Environment and Planning D: Society and Space*, 16, pp. 423–37.

Hartwick, E. (2000) Towards a geographical politics of consumption. *Environment and Planning A*, 32, pp. 1177–92.

Harrison, T. (1981) *A Kumquat for John Keats*. Newcastle: Bloodaxe.

Harvey, D. (1990) Between space and time: reflections on the geographical imagination. *Annals, Association of American Geographers*, 80, pp. 418–34.

Heal, C. and Allsop, M. (1986) *Queer Gear: How to Buy and Cook Exotic Fruits*. London: Century Hutchinson.

Henderson, D. (1992) Exotic produce: the changing market in the UK. *British Food Journal*, 94, pp. 19–24.

hooks, b. (1992) *Black Looks: Race and Representation*. London: Turnaround.

Hulme, P. (1986) *Colonial Encounters: Europe and the Native Caribbean 1492–1797*. London: Routledge.

Hulme, P. (1990) The spontaneous hand of nature: savagery, colonialism, and the enlightenment. In Hulme, P. and Jordanova, L. (eds), *The Enlightenment and its Shadows*. London: Routledge.

James, C. L. R. (1938) *The Black Jacobins*. London: Alison and Busby.

Johns, L. and Stevenson, V. (1985) *Fruit for the Home and Garden*. London: Angus and Robertson.

Kale, S., McIntyre, R. and Weir, K. (1987) Marketing overseas tour packages to the youth segment: an empirical analysis. *Journal of Travel Research*, 25, pp. 20–4.

Klein, N. (2000) *No Logo*. London: Flamingo.

Konder, R. (1982) The Carmen Miranda museum: Brazilian bombshell still box office in Rio. *Americas*, 34(5), pp. 17–21.

las Casas, B. de (1992) *A Short Account of the Destruction of the Indies (1552)*. Harmondsworth: Penguin.

Lindsay, A. (1989) *Sleeve Notes to Beleza Tropicale*. New York: Fly/Sire Records.

Litwack, D. (1989) 1989 Produce operations review. *Supermarket Business*, 44, pp. 41–50.

López, A. (1993) Are all Latins from Manhattan? Hollywood, ethnography and cultural colonialism. In King, J., López, A. and Alvarado, M. (eds), *Mediating Two Worlds: Cinematic Encounters in the Americas*. London: British Film Institute.

Mackintosh, M. (1977) Fruit and vegetables as an international commodity. *Food Policy*, November, pp. 277–92.

Manthorne, K. (1984) The quest for a tropical paradise: palm tree as fact and symbol in Latin American landscape imagery, 1850–1875. *Art Journal*, 44, pp. 374–82.

Market News Service (1989) *Horticultural Products Newsletter: Papaya (June)*. Geneva: ITC UNCTAD/GATT.

Mason, P. (1990) *Deconstructing America*. London: Routledge.

May, J. (1996a) 'A little taste of something more exotic': the imaginative geographies of everyday life. *Geography*, 81, pp. 57–64.

May, J. (1996b) Globalisation and the politics of place: place and identity in an inner London neighbourhood. *Transactions of the Institute of British Geographers*, 21, pp. 194–215.

Montrose, L. (1991) The work of gender in the discourse of discovery. *Representations*, 33, pp. 1–41.

Moore, K. and Utterback, L. (1996) Can passionfruit go mainstream? *International Horticulture* September–October (www.fintrac.com/gain/marketstats/sepoct96/sept1.htm).

MSI (1988) *Fruit UK, Marketing Database*. Surrey: Marketing Strategies for Industry.

Murray, W. (1971) The return of Busby Berkeley. In McClure, A. (ed.), *The Movies: An American Idiom*. Rutherford: Fairleigh Dickinson University Press.

Nash, C. (2000) Historical geographies of modernity. In Graham, B. and Nash, C. (eds), *Modern Historical Geographies*. Harlow: Prentice Hall.

Patterson, O. (1969) *The Sociology of Slavery*. London: Associated University.

Pieterse, J. N. (1992) *White on Black: Images of Africa and Blacks in Western Popular Culture*. New Haven: Yale University Press.

Porter, R. (1990) The exotic as erotic: Captain Cook in Tahiti. In Rousseau, G. and Porter, R. (eds), *Exoticism in the Enlightenment*. Manchester: Manchester University Press.

Pratt, M. L. (1992) *Imperial Eyes: Travel Writing and Transculturation*. London: Routledge.

Rand, E. (1995) *Barbie's Queer Accessories*. Durham, NC: Duke University Press.

Richard, P. (1991) 1492: The violence of God and the future of Christianity. In Boff, L. and Elizondo, V. (eds), *1492–1992: The Voice of the Victims*. London: SMI Press.

Roberts, S. (1993) 'The lady in the tutti-frutti hat': Carmen Miranda, a spectacle of ethnicity. *Cinema Journal*, 32, pp. 3–23.

Rowe, W. and Schelling, V. (1991) *Memory and Modernity*. London: Verso.

Rubin, M. (1993) *Showstoppers: Busby Berkeley and the Tradition of Spectacle*. New York: Columbia University Press.

Seed, P. (1993) 'Are these not also men?': the Indians' humanity and capacity for Spanish civilisation. *Journal of Latin American Studies*, 25, pp. 629–52.

Simon, R. (1994) Forms of insurgency in the production of popular memories: the Columbus quincentenary and the pedagogy of counter–commemoration. In Giroux, H. and McLaren, P. (eds), *Between Borders: Pedagogy and the Politics of Cultural Studies*. London: Routledge.

Springhall, J. (1970) Lord Meath, youth and empire. *Journal of Contemporary History*, 4, pp. 97–111.

Taussig, M. (1992) *The Nervous System*. London: Routledge.

Todorov, T. (1984) *The Conquest of America*. New York: Harper and Row.

Touissant-Samat, M. (1992) The tradition of fruits. In Touissant-Samat, M. (ed.), *A History of Food*. Oxford: Blackwell.

Van Den Abbeele, G. (1984) Utopian sexuality and its discontents. *L'Esprit Créteur*, 24, pp. 43–52.

Walker, R. Hardstaff, S. and Crow, D. (1994) South Atlantic souvenirs and trouble: the trophy cabinet. In Walker, R., Hardstaff, S. and Crow, D. (eds), *Trophies of Empire*. Liverpool: Bluecoat Gallery.

Watts, M. (1993) Living under contract: work, production politics and the manufacture of discontent in peasant societies. In Pred, A. and Watts, M. (eds), *Reworking Modernity: Capitalisms and Symbolic Discontent*. New Brunswick: Rutgers University Press.

Watts, M. (1999) Commodities. In Cloke, P., Crang, P. and Goodwin, M. (eds), *Introducing Human Geographies*. London: Arnold.

Willis, S. (1991) *A Primer for Daily Life*. London: Routledge.

Part IV

Ethical commodity chains and the politics of consumption

10 Unravelling fashion's commodity chains

Louise Crewe

Standardisation has become, over the years, a real danger [...] What's the point of going to town centre B if you're going to get a carbon copy of what you had in town centre A? It's more and more likely that there will always be the same Boots and the same Next.

(Burt, cited in Taylor, 2002)

If Nike Town and the other superstores are the glittering new gateways to branded dreamworlds, then the Cavite Export Processing Zone, located ninety miles south of Manila in the town of Rosario, is the branding broom closet.

(Klein, 2000, 202)

I finish off the speech I'm giving at the World Trade Organization (WTO) protest meeting on the British Airways flight from Heathrow to Seattle. A usually dull journey is invigorating because the flight is an extraordinary experience in itself: everyone is on the way to do battle in Seattle. The plane is packed with Australian activists, anti-logging campaigners, War on Want members...The NGOs are out in force and they alone will be standing up for billions of unrepresented people and will be putting their case to the representatives of the most powerful corporations in the world.

(Roddick, 2000, 3)

Introduction

The vignettes above encapsulate three themes central to the chapter, namely a current crisis facing high street fashion retailers; shifting global supply relations pursued by fashion companies; and the possibilities which these issues raise for a politics of consumption and questions of corporate responsibility and ethical trade.[1] In the following discussion I first consider transformations in the fashion retail system within western industrial economies. Prompted by intensified competition, market saturation, rising costs and shifting consumption preferences, high profile fashion retailers such as Marks and Spencer (M&S) and Next in the United Kingdom; Puma and Hugo Boss in Germany; and Donna Karan, The Gap, Ralph Lauren, Timberland and Nike in the United States are reconfiguring their supply relations. This is having significant impacts on the geographies of fashion production.

Second, I suggest that emerging geographies of fashion raise important questions about the management, practice and conceptualization of global supply chain relations in the fashion industry. In particular I argue that the spaces of fashion production and consumption are inextricably linked. Connections between nodes are complex and difficult to untangle; and the power relations which underpin them are mobile and diffuse. Fashion manufacturers such as M&S and the Burton Group in the United Kingdom have no factories at all; designers such as Calvin Klein are not themselves the producers of the items that bear their name; and retailers such as Benetton do not actually own any shops. In a world saturated with advertising and branding, the image of a fashion house or designer must be carefully manufactured and orchestrated across both economic and cultural sites.

Although a commodity chain approach offers one means of tracing the complex geographies of image manipulation across distant and distinct sites of production and consumption, it is not fully able to theorize the fashion system. Originally developed by Gereffi (1994) and others within a political economy perspective, the global commodity chain (GCC) centrally distinguishes between producer-driven and buyer-driven commodity chains. The utility of the approach is its focus on territory, governance and institutional frameworks which facilitates the theorization of power networks, barriers to entry, sourcing relations and supply chain management. Using insights from the GCC approach, one could argue that the fashion system is a buyer-driven chain in which producers (whose barriers to entry are low) are subordinated to key actors who control design, marketing, advertising, branding and retail nodes in the chain (see Crewe and Davenport, 1992). However, the distinction between producer and buyer-driven chains fails to take account of chains where a number of nodes are assigned key agency, such that there might be a number of 'drivers' in any one GCC (Raikes *et al.*, 2000). Furthermore, GCC approaches take little account of other non-buyer or producer drivers, such as governments or NGOs who potentially may influence particular parts of the chain. The issue of regulation is particularly significant. As Raikes *et al.* (2000, 400) have argued, the 'concentric ring' structure observed in global textile and apparel trades in large part has been shaped by shifting regulatory environments such as the Multifibre Arrangement and the North American Free Trade Agreement (NAFTA). Similarly, minimum wage legislation and ethical codes legislation have the potential to transform the spaces in which commodity chains operate and the practices through which they are (re)produced. GCC approaches also fail sufficiently to examine the consumption end of the chain, although work within the broader 'systems of provision' tradition is more helpful in this respect (Fine, 1995, 1993; Fine and Leopold, 1993). Understandings of the intersecting dynamics of production and consumption spaces and practices continue to develop, although several conceptual challenges still remain. These include a need to understand the ways in which nodes within particular chains intersect and diverge in time and space; to locate shifting power relations across particular chains; to explore potentialities of shifting regulatory regimes for reshaping commodity chains and to trace the discourses, knowledges and representations through which commodity chains are constructed.

A third key focus of the chapter is upon attempts to regulate the global fashion industry. The act of consumption increasingly is invested with strong political overtones and the repercussions of consumption are becoming increasingly evident both in terms of the environmentally devastating effects of over-consumption and in terms of the pernicious nature of labour processes central to the low-cost production of commodities such as clothes, toys and consumer electronics (Abrams and Astill, 2001; Johns and Vural, 2000; Klein, 2000; Roddick, 2000; Ross, 1997). Through unravelling the spaces and practices of production and consumption along fashion's commodity chains, the chapter seeks to open up debate about the potential development of ethical systems of provision and consumption (see also Hartwick, 2000). I explore connections between commodity culture, identity and political participation in order to illuminate the possibilities for a more radical consumer politics which can begin to link the signified to the material realities of fashion. While there is evidence of growing concern about corporate responsibility, ethical trade and responsible consumption, the extent to which emergent corporate codes of conduct and sustainable trade initiatives can ever regulate the operations of private for-profit multinational firms is debatable.

Fashion victims: concentration and crisis on the high street

The organizational geographies of fashion supply chains have been well rehearsed: many commentators have documented the growing power held by large retail capital during the 1980s (Crewe and Davenport, 1992; Entwistle, 2000; Phizacklea, 1990). This period saw a growth in the middle reaches of the fashion market with retailers such as Next, Principles and M&S in the United Kingdom, and The Gap in the United States converging to supply the mass fashion market with affordable, coordinated ranges of business and casual wear (Crewe and Davenport, 1992; Crewe and Lowe, 1995). Concentration and near oligopoly was the characteristic mode of business organization in both Britain and the United States. In the United Kingdom during the early 1990s, almost 40 per cent of all sales derived from just six retail organizations. Brand concentration is an increasingly significant issue in the United States, where 30 per cent of clothing sales are accounted for by twenty companies including Liz Claiborne, Fruit of the Loom, Victoria's Secret and Oshkosh B'Gosh (NACLA, 1997). By 1998 the top ten names in fashion in the United States commanded 47 per cent of all sales (Anon, 1999).

Brand and store concentration has a series of implications for the global organization of fashion commodity chains, with the preferred supplier model emerging as one way of understanding shifting supply chain relations during the 1980s and 1990s (Crewe and Davenport, 1992). Particularly in the United Kingdom, the existence of a concentrated retail sector with increasingly stringent production requirements pushed certain sections of the clothing industry into a more competitive mould. The adoption of the preferred supplier model guaranteed large orders, close and long-term working relationships with retailers and consistent demand for UK suppliers. M&S, in particular, treated favoured suppliers well and ensured skilled employment for large numbers of designers and machinists across Britain's regions. However, since the early 1990s economic conditions in Europe and North America have become less favourable and the retail sector has faced rising costs and increased competition.

A key change to the structure of clothing retailing in recent years has been a marked decrease in concentration. In the UK concentration indices fell from 1996 onwards as middle-market retailers faced intensifying competition and stagnating demand. Retail polarization also poses particular threats to middle-market retailers, as strong growth within both upper and lower reaches of the market undermines continued mass middle-market saturation by large multiples. Discounters such as Matalan, New Look and Mark One have redefined the lower end of the UK clothing market and increasingly are offering fashionability and value. Sales in 'discount' sectors grew by 17 per cent in 1999 whereas the clothing market as a whole expanded by only 0.8 per cent – a trend which is predicted to continue (Verdict, 2000).[2] As price becomes an increasingly significant element in supply chain management strategies, sourcing chains and supply relations will be altered accordingly.

More expensive and design-led clothing retailers such as Jigsaw, Hobbs, French Connection and Whistles also are outperforming companies in the middle ranges of the market. French Connection saw a 20 per cent increase in profits during 1999–2000 and predicted a further 20 per cent increase for the following season in both the United Kingdom and the United States (Just-Style, 2000). While the market share captured by the 'big six'[3] retailers had been increasing steadily since the early 1980s, these firms began to lose ground from the mid-1990s onwards (see Table 10.1). By 1999 the largest six clothing retail organizations accounted for only 32.5 per cent of total sales. Arcadia axed 3,500 jobs in 2000 and M&S announced 4,390 job losses in 2001 – despite paying £20 million to management consultants Andersen in a bid to improve their ailing business. The Dutch retail chain C&A has withdrawn from the UK market entirely. At the same time new entrants such as the Spanish chains Zara and Mango and the Swedish chain Hennes and Mauritz (H&M) are making significant inroads into the middle reaches of the fashion market.

The crisis facing mass-market retailers in the United Kingdom can be traced to problems resulting from overcapacity, price deflation, changing consumption preferences and rising competition from domestic design-led retailers, overseas-based chains and domestic discounters. Although the next four years will see UK clothing and footwear retailers open another 10 million square feet of retail space, it has been argued that the middle market can no longer support all participants.[4] In addition, long lease arrangements and prohibitive exit costs have made it difficult for many retailers to restructure operations and/or close stores. Falling profits are one indicator of the problems facing middle market fashion retailers. In 1999, for example, Arcadia's profits fell by 38 per cent and sales at BhS declined by 9 per cent.

Table 10.1 Estimated market shares of clothing sales by major UK retailers, 1996–99

Retailer	Market share in %	
	1996	*1999*
M&S	16.5	11
Arcadia (Burton Group)	10	7.0
Storehouse	6.5	5.5
JIL (Sears)	5.3	4.5
Next	3.2	3.5
C&A	4	0
Total	46.2	31.5

Sources: Key Note Report 1996, Verdict Market Research Report 1999, CAFOD 1998.

Notes
Arcadia Group (formerly Burton Group) includes Dorothy Perkins, Top Shop, Top Man, Burton, Evans, Principles, Racing Green and Hawkshead.
Storehouse Group includes BhS, Mothercare, Blazer and Children's World.
JIL (formerly Sears) includes Adams Childrenswear, Miss Selfridge, Wallis, Warehouse and Richards.

However, it is M&S who exemplify the difficulties facing middle-market clothing retailers most acutely. Sales and market share have been falling steadily since the mid-1990s and the company culled one-quarter of its top management during 1999. Despite ongoing rebranding and marketing endeavours the group continues to be besieged by derogatory press reports, falling share prices and waning consumer confidence and respect. Although M&S continued to receive mention within market research surveys in the late 1990s, comments were made almost exclusively by customers in the over fifty-five age bracket (Verdict, 1999). Responding to difficulties both with profitability and their overall image on the high street, M&S sought to cut prices by an average of 10–15 per cent during 2000. Store interiors were revamped and product ranges updated for a more design-conscious consumer (Finch, 2001). For the first time in its history, M&S launched television and print media advertising campaigns. However the retail recessionary pressures which precipitated these developments also engendered a reorganization of management practices and supply relations, and it is to these that I now turn.

A jerky, deranged sort of dance:[5] disentangling fashion's spatialities

A volatile retail environment has caused retailers to rethink supply chain relations and to wield corporate power in unexpected ways. During the 1980s the control that clothing retailers exerted over supplier firms seemed unshakeable: few could have predicted the scale of the shift away from domestic suppliers in the following decade. However middle-market retailers now have become increasingly less concerned about the origins of clothing providing it matches price points and quality standards. As conditions on the high street tighten, a range of fashion retailers are re-evaluating their commitment to domestic sourcing and are reordering their supplier relations into ever more tangled webs of global connection.

Next, for example, bought 80 per cent of its merchandise from UK suppliers in the mid-1980s, but by the late 1990s this share had fallen to 40 per cent. As I have indicated, M&S traditionally put long-term cooperative relationships with suppliers and customer service at the heart of its policy. However more recently the group has reneged on its 'Buy British' policy and has pursued a more aggressive strategy of international sourcing, with predictable outcomes for British clothing firms. In the 1980s the store sourced 90 per cent of its clothes from the United Kingdom. By the early 1990s the figure reached 60 per cent and has been falling steadily ever since, reaching 50 per cent in 1999 with a plan to reduce this to 30 per cent domestic sourcing by 2002. The most recent trend has been for M&S to persuade suppliers to source more of their products overseas, allowing the retailer to improve margins without passing on higher costs to customers. In instituting fundamental changes to buying policies and supply-chain management strategies, including the termination of long-standing contractual arrangements and the severing of close relations with suppliers, M&S has exacerbated crises facing the UK clothing industry. The Scottish group William Baird, for example, had been involved in the design, manufacture and sale of men's, women's and childrenswear to M&S for thirty years. M&S severed this deal without warning in October 1999, a move estimated to cost 8,000 Baird jobs. Ongoing company strategy sought to locate the majority of production overseas by 2000: M&S childrenswear, for example, is now sourced exclusively from factories in China, Sri Lanka and Indonesia where labour cost savings are substantial (see Table 10.2). Courtaulds, who are the largest UK producer of lingerie and underwear and sell a third of their output to M&S are currently developing a worldwide network of manufacturing operations and are already supplying garments from lower cost businesses in China, Indonesia, Thailand, Morocco, Tunisia, Turkey and the Philippines. It has been estimated that 10,000 UK clothing manufacturing jobs could be lost by M&S's change of sourcing strategy (Oxborrow, 2000), and that 80,000 clothing jobs were lost in total in the United Kingdom between 1998 and 1999.

A parallel situation is evident in the United States where a reported average of 41,416 textile and apparel jobs have been lost each year since 1979 (Ross, 1997, 82). In the city of New York alone 6,000 apparel jobs were lost in 1994 (Wark, 1997, 238). The local effects of such spatial-switching of sourcing strategies are particularly problematic in regions such as the East Midlands and the North East in the United Kingdom, and in US conurbations such as New York where clothing has traditionally been a large employer, particularly of women. In both countries, shifting supply relations have led to a marked increase in clothing

Table 10.2 Global variations in clothing manufacturing wage rates

Country	Hourly compensation (wages and benefits), in $
UK	7.00
US	8.00
Mexico	0.85
Thailand	0.38–0.48
Dominican Republic	1.15
Malaysia	1.15
Indonesia	0.15
Sri Lanka	0.13
China	0.12–0.125
Vietnam	0.12
Bangladesh	0.07

Source: Kurt Salmon Associates/Sweatshop Watch 1995.

imports: 60 per cent of clothing bought by UK consumers (Verdict, 1998) and 66 per cent of apparel sales in the United States (Figuera, 1997) are accounted for by imports.

Given global variations in wage rates, and in part as a result of neo-liberal economic restructuring strategies and free trade agreements, large retailers are able to relocate in low labour cost countries seemingly regardless of the impacts their decisions will have on either domestic clothing industries or the predominantly young women who are exploited under sweatshop conditions for what typically amounts to less than a living wage. While the production histories of many fashion garments are obscured by lax labelling legislation and seductive advertising campaigns, a glance behind the glamour of the consumption space reveals a tenuously connected tissue of sites which criss-cross the globe in quite spectacular ways.

A pair of Lee Cooper jeans on sale at a large discount store in a provincial British city, for example, belies the conditions and spaces of their production (Abrams and Astill, 2001). Displayed under a large sign announcing 'Famous brands for £19.95', the retail site is one stop on a 40,000 mile journey (ibid., 2). The jeans arrived

> in a van that came up the A12 from Lee Cooper's warehouse at Staples Corner, just at the bottom of the M1 in North London...Before that they came through the Channel Tunnel in a lorry from similar warehouses in Amiens, France and before that, by boat and train from Tunis in Tunisia.... But this factory is not the beginning for our pair of jeans. In one sense it's the end. The destination.

> (ibid., 2–3)

The journey traced in Figure 10.1 and Plate 10.1 of the jeans is in fact only a partial story of the life of the commodity. The jeans will have been designed in the United States, advertised globally, retailed in stores across Europe and the United States and will end up in a wardrobe, on someone's body, ready to enter another series of journeys in their biography. Fashion's commodity chains are labyrinthine, convoluted and tortuous. And while certain sites – design houses, flagship stores, advertisements – are thrown into high relief, other spaces such as cut-make-and trim plants, and fibre and fabric sourcing zones are hidden from view. It is to the politics of such concealed spatialities that I now turn.

'You're a damned fool if you own it':[6] sweatshops behind the swoosh

Clothing consumption has been rising steadily since the mid-1980s, with sportswear showing particularly marked increases. The sportswear market is currently valued at £3.65 billion and sportswear accounted for 14.6 per cent of all clothing and footwear purchases in the United Kingdom in 1999 (Key Note, 2000). The market for replica football shirts is particularly strong, standing at £150 million in the late 1990s (Key Note, 2000), with each shirt costing in the region of £50.00. Training shoes too enjoyed heady growth throughout the 1990s, with customers prepared to pay in excess of £100.00 several times per year for new

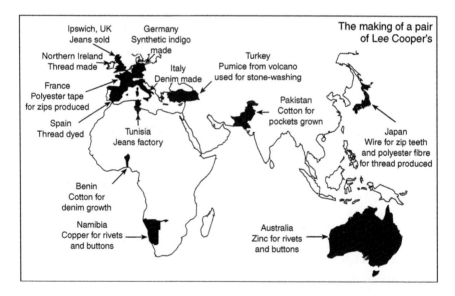

Figure 10.1 The making of a pair of Lee Cooper jeans.

Source: *The Guardian*, G2, 29.05.01: P3.

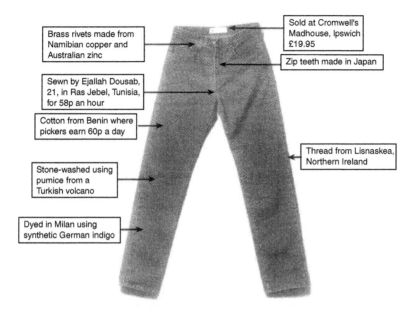

Brass rivets made from Namibian copper and Australian zinc

Sold at Cromwell's Madhouse, Ipswich £19.95

Zip teeth made in Japan

Sewn by Ejallah Dousab, 21, in Ras Jebel, Tunisia, for 58p an hour

Cotton from Benin where pickers earn 60p a day

Thread from Lisnaskea, Northern Ireland

Stone-washed using pumice from a Turkish volcano

Dyed in Milan using synthetic German indigo

Plate 10.1 There's more to a pair of jeans than you thought.
Source: *The Guardian* G2, 29.05.01: P1.

statement trainers. The sports footwear market is predicted to increase to £1.5 billion by 2004 (Key Note, 2000).

However, the contribution to the domestic market from indigenously sourced manufacturing in both the general clothing market and the sportswear market is deteriorating. Large retail buyers are able to take advantage of a global pool of workers whose wages differ for historical, institutional, political and economic reasons. Companies traditionally satisfied with a 100 per cent mark-up between the cost of factory production and the retail price now scour the globe for factories in which production is so inexpensive that the mark-up is closer to 400 per cent (Klein, 2000: 197). Workers in a Nike factory in Bangladesh, for example, earn less than 0.5 per cent of a jacket's retail value (CAFOD, 1998).

However few consumers are aware of the origins of their clothes. Country of origin labelling is not a legal requirement in Britain and neither the positioning of fashion products in the flagship stores of global cities such as Nike Town and The Disney Store in London and New York, nor glamorous advertising campaigns reveal stories about the hands that cut, sew and pack such commodities or about the spaces in which production takes place. On first reading, it would appear that retailers' endeavours to create and sustain glamorous corporate images at retail and consumption sites and to mask less obvious and less seductive material conditions at sites of production have been a success. In a classic example of the international division of labour, footloose multinationals in a borderless world have reshaped geographies of fashion production, with predictable results:

> money without borders leads to sweatshop exploitation of the world's poorest. Industry after industry seems perfectly happy to use sweatshops and the globe is quickly becoming a playground for those who can move capital and projects quickly from place to place. When business can roam from country to country with few restrictions in its search for the lowest wages, the loosest environmental regulations and the most docile and desperate workers then the destruction of livelihoods, cultures and environments can be enormous.
>
> (Roddick, 2000, 7)

Neo-liberal regulatory policies and discursive constructions of modernization by governments and multinational firms are serving to reshape fashion commodity networks. Key destinations for multinational fashion firms are Export Processing Zones (EPZs)[7] where legislation is minimal and tax exemptions bountiful. Comparatively cheap labour costs and liberal regulatory environments further enhance perceived competitive advantages offered by EPZs, while the competitive edge conferred by proximity to markets offers significant advantages to low cost labour locations on the periphery of the large consumer markets of western Europe and the United States. Trade agreements such as NAFTA have resulted in a marked shift of fashion production away from the traditional urban centres of Toronto, Los Angeles and New York and towards assembly operations in Mexico, Central America and the Caribbean. Between 1995 and 1998 Mexican apparel exports to the United States increased by 119 per cent, reaching US$6 billion by the end of the period (Kessler, 1999). Worldwide, the ILO estimates that 1,000 EPZs employ some 27 million workers (Klein, 2000, 205).

However labour conditions and work practices in many EPZs increasingly are attracting public and media scrutiny. Chinese Special Economic Zones are a particularly prominent example of the pernicious work environments associated with garment production. From an insignificant global fashion producer in the late 1980s, China now has become the world's largest garment exporter. In the last quarter of 1998, 38 per cent of British garment imports came from China and Hong Kong, while in 1999 China accounted for one-sixth of the world garment trade (Just-Style, 2000). Estimates suggest that over four million garment workers are employed in China's 124 SEZs. In 1999 US$1.46 billion was invested in China's clothing SEZs, transforming the former farm fields of Dongguan (Ehrlich, 2000). Nike alone employs 70,000 workers and has attracted serious criticisms for employment conditions in Chinese factories. The National Labour Committee report *Made in China* details serious abuses of human rights, including 11–12 hour shift patterns for 6 days per week with base wages of US 20 cents per hour (NLC, 2000), while *Sweatshops behind the swoosh* similarly reports illegally long hours and a ban on unions in Nike factories (UNITE, 2000).

Allegations of race, age and gender discrimination in zone economies abound: the ILO estimates that at least 10,000 children work in the Bangladeshi garment industry (Spielberg, 1997, 113) and the National Garment Workers Federation suggests that sexual abuse is commonplace.[8] Profit advantages for investors are underscored in publicity material: 'Bangladesh offers the most inexpensive and productive labour force. Law forbids formation of trade unions and strikes are illegal' (cited in Asia Monitor Resource Centre, 1998). Drawing upon long-standing constructions of women's compliance, nimble fingers and need only for 'pin money', such advertising reproduces unequal gender roles and justifies exploitative work relations (Elson, 1984; Fernandez-Kelly, 1983; Fuentes and Ehrenrich, 1983; Phizacklea, 1990). There is a surprising persistence of essentialized myths about women and work which are discursively constructed and reproduced across zone economies, both inside and outside the factory gate. In Mexico, representations of maquiladora women as 'docile, submissive and tradition bound worker[s] who will only be suited to positions of least prestige in the workplace' result in their exclusion from training programmes critical to the skilling process (Wright, 1997, 279). Young and unorganized women often are objectified and sexualized by Maquila managers 'whose practices, representations and imaginations have profound repercussions for the lives of other(s)' Salzinger (2000, 80).

There are dangers, of course, in uncritically rehearsing practices of gender and age discrimination, and a number of studies have turned their attention to the formation of gendered categories in order to move beyond simply recounting and lamenting the fate of women workers (Berndt, 2001; Catanzarite Strober, 1993; Ong, 1997; Salzinger, 2000; Wright, 1997, 1999). Divisions and dualities between men and women; between compliance and choice; between export plants and others are cleaved through with a number of striations that make broad generalizations difficult to sustain. For many women in the South, entry into the paid labour market offers control and independence (Ong, 1987), and for many poor families the clothing industry provides much needed income. Set against the experience of working for domestically owned firms, workers may obtain marginally improved terms and conditions of employment from subcontractors to large multinationals and may obtain status from producing international brands (Cox Edwards, 1996; Pangestu and Hendytio, 1997; Smyth and Grijns, 1997; Tjandraningshi, 1995). Such complexities reveal difficulties in straightforwardly making political connections along commodity chains.

Brand capitalism, ethical trade and the politics of consumption

As a result of media exposure and NGO campaigns, multinational corporations across a range of sectors including fashion and footwear have been forced to reconsider the ways in which they do business, particularly in relation to environmental and labour practices (Hale, 2000; Hartwick, 2000; Johns and Vural, 2000). This has resulted in endeavours on the part of large retail capital to reorder global labour practices and to respond to increasingly vociferous calls for more sustainable and ethical trading relations.[9] The use of child labour in supply chains increasingly is acknowledged as one of the greatest risks to a company's reputation and its elimination has become part of broader moves toward greater corporate social responsibility.

In the United Kingdom and the United States the ethical trading agenda accelerated in the late 1990s as the deleterious effects of off-shore production became apparent. A 1997 Market and Opinion Research International (MORI) poll suggested that 92 per cent of British consumers believed that companies should have minimum standards for 'third world' suppliers (CAFOD, 1998). Following the potentially damaging revelation that jeans were being made by Chinese prison labour on the island of Saipan, Levi Strauss immediately drew up a code of conduct for all suppliers. Nike too developed code of conduct arrangements at an early stage – although their campaigns were not always to good effect as we shall see below. The majority of large fashion retailers who source commodities overseas now have established some form of code of conduct, although the scope and scale of initiatives is highly variable. Some codes have been designed to apply at the level of the single retailer (Levi's in the Dominican Republic, e.g.), while others are national sectoral codes negotiated between trades unions and employers, as in the case of Costa Rica. Codes such as those regulating the Hong Kong toy industry have been introduced in response to external pressure (Coalition for the Charter on the Safe Production of Toys, 1996).

The most comprehensive codes are based upon core labour standards enshrined in ILO conventions; and those with the greatest potential impact are formalized multi-agency codes such as codes developed by the International Confederation of Free Trades Unions (ICFTU). The ICFTU has sought to pressure national governments to ratify and fully apply ILO Conventions 138 and 182 on minimum age requirements and child labour. What is particularly significant about the most recent wave of code activity is a shift in the conceptualization of the appropriate sites and spaces for intervention. Rather than focusing entirely on sites of production as earlier labour movements had done, attention increasingly is targeted at the entire commodity chain and specifically at sites of consumption.

It has long been acknowledged that fashion is a feminized commodity chain, but increasingly it has become clear that concerns affect women across all sites – from consumers troubled by questions of origin and price, to retail workers faced with casualized and intensified work practices, to skilled machinists in the urban economies of the United States and United Kingdom whose jobs are being eroded, through to young women working for below-subsistence wages in global EPZs. There are connections to be made across the chain – involving questions of gender, identity, performance and presentation – which are specific to fashion and relate to a 'fashion chain logic' (Leslie and Reimer, 1999, 408). One of the earliest attempts to expose and improve working conditions for women in the global garment chain came in the form of the Stop Sweatshops Campaign organized by the Union of Needletrades, Industrial and Textile

Employees (Johns and Vural, 2000). Working in coalition with consumer, human rights and labour organizations, UNITE's campaign was one of the first to direct attention towards fashion retailers. It also sought to illuminate the plight of women garment workers in a range of locations rather than privileging one group of workers at the expense of others (Johns and Vural, 2000). A defining feature of the campaign was a focus on consumers as agents for change, acknowledging that spaces of consumption are highly significant in attempts to intervene in commodity chains.

More recently the Ethical Trading Initiative (ETI) has been launched in the United Kingdom as a pilot scheme to improve labour practices across a range of sectors, primarily food and fashion. Having developed from Fair Trade Campaigns run by British aid organizations, the ETI is based around a multi-agency grouping including companies, NGOs and trades unions who have devised a base code for ethical labour practices (see Appendix). The code includes the following components:

1. Employment is freely chosen.
2. The right to collective bargaining is respected.
3. Working conditions are safe and hygienic.
4. Child labour shall not be used and employment policies must conform to the ILO standards.
5. Living wages are paid and all workers must be supplied with written information about wages and conditions.
6. Working hours are not excessive. All workers will be given one day off in seven and should not work in excess of 48 hours per week. Overtime is voluntary.
7. No discrimination is practised.

Organizations involved in the ETI have considerable economic power. Participating firms have combined turnover of £50 billion and subcontract to more than five million factories, farms and plantations; NGOs have in excess of one million members; and union delegates represent 127 million workers worldwide (DFID, 1998). However questions remain about the potential of the ETI to effect real changes in the organization of the fashion commodity chain, in part because of limited dialogue between code implementers and the distant production units which the code is endeavouring to regulate. The complexity of international sub-contracting networks and the intense competitive pressures facing manufacturers creates practical problems: when the retailer C&A first set up a monitoring system it took four years to determine which factories were making their clothes (quoted by Williams, Head of Corporate Communications at C&A, 1998).

Nike provides a good example of the problems and pitfalls associated with an audit culture. The corporation has agreed to endorse the Code of Conduct of the Coalition for Environmentally Responsible Economics (CERES) which is aimed at creating a framework for dialogue and negotiation across global supply chains.[10] Having also introduced a programme of external audits by Price Waterhouse Coopers (PwC) called Transparency 101, Nike argue that they now have the most comprehensive factory-monitoring systems available. Yet the company has published details only of a few selected factories whose names and addresses remain confidential, and PwC audits of Nike factories in China, Korea and Cambodia have been viewed as flawed. Speaking about a 1998 US Fair Labour Association Audit of Nike factories Ehrilch (2000) argues that 'auditors acknowledge that their visits are snapshots at best and rarely carry an element of surprise. For example, when SA8000 auditors visit Chinese factories their inspections are carried out every six months, like clockwork'. Such comments cast doubt on the potential success of audits: assessors working for a for-profit firm selected by and accountable to Nike are unlikely to develop the trust of employees necessary for them to engage in genuinely open dialogue. Although Nike holds up monitoring processes as evidence of best practice, labour conditions in many subcontracting factories flout international conventions, with children often working 7 days per week for up to 16 hours per day (Kenyon, 2000).

A widely circulated report of an email correspondence with Nike reveals the disjuncture between the public face of the corporation and the less visible geographies of global fashion production. The example reveals the interconnectedness between spatially and organizationally distant nodes within fashion's commodity chain and demonstrates the difficulties faced by large retailers whose corporate image-making processes are being undermined by a growing body of consumers intent upon bringing the entire commodity chain into view. Having selected the word 'sweatshop' to be stitched onto a 'personalized' pair of Nike shoes, Jonah Peretti received the following correspondence:

> From: 'Personalize, NIKE Id' <nikeid_personalize@nike.com>
> Your NIKE iD order was cancelled for one or more of the following reasons: 1) Your Personal iD contains another party's trademark or other intellectual property. 2) Your Personal iD contains the name of an athlete or team we do not have the legal right to use. 3) Your Personal iD was left blank. Did you not want any personalization? 4) Your Personal iD contains profanity or inappropriate slang.

> From: 'Jonah H. Peretti' <peretti@media.mit.edu>
> My order was cancelled but my personal NIKE iD does not violate any of the criteria outlined in your message. The Personal iD on my custom ZOOM XC USA running shoes was the word 'sweatshop.' Sweatshop is not: 1) another's party's trademark, 2) the name of an athlete, 3) blank, or 4) profanity. I cho[o]se the iD because I wanted to remember the toil and labour of the children that made my shoes. Could you please ship them to me immediately.

> From: 'Personalize, NIKE iD' <nikeid_personalize@nike.com>
> Your NIKE iD order was cancelled because the iD you have chosen contains, as stated in the previous e-mail correspondence, 'inappropriate slang'.

From: 'Jonah H. Peretti' <peretti@media.mit.edu>
Thank you for your quick response to my inquiry...Although I commend you for your prompt customer service, I disagree with the claim that my personal iD was inappropriate slang. After consulting Webster's Dictionary, I discovered that 'sweatshop' is in fact part of standard English, and not slang. The word means: 'a shop or factory in which workers are employed for long hours at low wages and under unhealthy conditions' and its origin dates from 1892. So my personal iD does meet the criteria detailed in your first email. Your web site advertises that the NIKE iD program is 'about freedom to choose and freedom to express who you are.' I share Nike's love of freedom and personal expression. The site also says that 'If you want it done right...build it yourself.' I was thrilled to be able to build my own shoes, and my personal iD was offered as a small token of appreciation for the sweatshop workers poised to help me realize my vision. I hope that you will value my freedom of expression and reconsider your decision to reject my order.

From: 'Personalize, NIKE iD' <nikeid_personalize@nike.com>
Regarding the rules for personalization it also states on the NIKE iD web site that 'Nike reserves the right to cancel any personal iD> up to 24 hours after it has been submitted'...With these rules in mind, we cannot accept your order as submitted.

From: 'Jonah H. Peretti' <peretti@media.mit.edu>
Thank you for the time and energy you have spent on my request. I have decided to order the shoes with a different iD, but I would like to make one small request. Could you please send me a color snapshot of the ten-year-old Vietnamese girl who makes my shoes?

<no response>

Source: crit-geog-forum@jiscmail.ac.uk, 08/03/01, item number 004351;
cited in *The Guardian*, 19 February 2001.

Auditing systems are, for the moment at least, failing to end labour abuses: formalized, one-off systems are unlikely to ever be an appropriate means of tackling systematic labour and environmental abuses on a global scale. Effective monitoring must be part of a broader, continuous process:

A Fair Trade Policy must be worth more than the paper it is written on. It is only too easy for industry to kid itself and its customers that everything in the garden is rosy. Our experiences over many years show that assurances from overseas agents and suppliers should be independently verified.

(Wendy Wrigley, Head of Labelling, The Co-op, DFID, 1998)

Conclusions

The preceding discussion has sought to emphasize the mutual dependence of production and consumption nodes within fashion's commodity network. Endeavours by large retail companies to promote fashionable and desirable brand names through flagship stores, new product launches, licensing deals and upscale advertising campaigns, while superficially concealing hidden labour processes at production sites, are increasingly being called into question. Consumers' geographical imaginations increasingly roam across the entire commodity chain, simultaneously bringing into view consumption and production sites, and the vision is increasingly intolerable. The spaces in which fashions are sewn are just as important as the spaces in which they are advertised, modelled, purchased and worn, resulting in increasingly vociferous demands for ethical systems of provision.

However despite a growing interest in more socially responsible supply chain management strategies, the prospects for labour relations in the globalized fashion industry remain bleak. As currently instituted, it is unlikely that corporate codes of conduct ever will be an appropriate means of regulating globalized commodity chains, particularly in view of the fact that workers are rarely consulted during the process of code design or implementation (Hale, 2000). Companies have expressed commitment to formalized codes of conduct but are still arguably engaged in little more than public relations exercises. Codes devised and implemented by head offices in the first world and involving little or no dialogue with workers are unlikely to effect real change. Without greater participation from export countries, codes as currently instituted will remain top-down and paternalistic. Developing a politics of consumption under such circumstances is a formidable task.

The encounter with the fashion commodity chain outlined in this chapter emphasizes two key issues currently neglected within existing GCC literatures. First, it is crucial not to focus exclusively on one moment in a commodity's life – whether its origin as a raw material, its production for market or its mediation in advertising. The tendency in existing literature is to run different moments together or to give undue prominence to one, such that production, mediation or reception becomes the 'determining instance' (Lury, 1996: 19). The danger is that a delicately balanced sequence of relationships across space and time – a 'deranged dance' (Abrams and Astill, 2001) – all too often become obscured and replaced by a simplistic set of reductions, ignoring the changes in form and meaning of commodities as they circulate through lives of production, promotion, sale, reception, use and reuse (Crewe and Gregson, 1998; Lury, 1996). The chapter has emphasized that it is the interdependence of nodes within the chain which is the key feature, and has suggested that the spatialities of chains and networks often are ignored within existing accounts.

Second, shifting regulatory regimes at all spatial scales also require greater theoretical attention. At the international level, as Raikes *et al.* (2000) so clearly point out, there seems to be an assumption that the 'deregulation' which characterizes the present period means no more that a reduction in regulation, whereas in fact, it more often than not means a shift in the type and form of regulation. This is important if we are to move towards a more theoretically inflected understanding of the shifting dynamics both within and across global commodity chains.

Appendix

List of ETI members

Companies

Arbor International	Anchor Seafood
ASDA	The Body Shop International
Chiquita International Brands	CWS/The Co-op
Debenhams Retail	Desmonds & Sons
Dewhirst Group	Fisher Foods
J. Sainsbury	Levi Strauss
Littlewoods	Lyons Seafoods
Lambert Howarth Global	M&W Mack
Madison Hosiery	Marks & Spencer
Monsoon	Mothercare
NEXT	Pentland Group
Premier Brands	Safeway Stores
Somerfield Stores	Tea Sourcing Partnership*
Tesco	

Trade unions

International Confederation of Free Trade Unions
International Textile, Garment and Leather Workers' Federation
International Union of Foodworkers
Trades Union Congress

Non-governmental organizations

Africa Now	Anti-Slavery International
CAFOD	Central American Women's
Christian Aid	Network
Labour & Society International	Fairtrade Foundation
Quaker Peace and Social Witness	Oxfam
Traidcraft Exchange	Save the Children
War on Want	TWIN
World Development Movement	Women Working Worldwide

Source: Adapted from
http://www.ethicaltrade.org/pub/members/list/main/index.shtml

Note
* The Tea Sourcing Partnership is an association of tea packers comprising Matthew Algie & Co., Brooke Bonde Tea Co., Finlay Beverages, Gala Coffee & Tea, DJ Miles & Co, Nambarrie Tea Co, Keith Spicer, Taylors of Harrogate, Tetley GB, R. Twining & Company and Williamson & Magor Co.

Notes

1 Blowfield (1999) defines ethical trade as the sourcing of products from companies guaranteeing core labour and human rights standards to their workforce.
2 The discount giants Wal-Mart and K-Mart in the United States present similar challenges for middle-market retailers and price competition at the lower reaches of the market is having a systemic negative effect on profit margins. As in the UK case, the middle market effectively is being squeezed by the price-led discounters at the bottom end and the design-led fashion houses at the upper reaches of the market.
3 The 'big six' retailers are the most powerful groups who capture the largest shares of clothing sales. Tracking their precise market share has become increasingly difficult due to dramatic acquisition, take-over and exit activity. While the 'big six' were consistently M&S, The Burton Group, Sears, Storehouse, C&A and Next throughout the 1980s and 1990s, this profile shifted considerably when The Arcadia Group (formerly the Burton Group, which included Dorothy Perkins, Top Shop, Top Man, Burton, Evans, Principles, Racing Green, Hawkshead) bought the Sears Group (including Miss Selfridge, Wallis, Warehouse and Richards) and BhS. C&A, meanwhile, have ceased trading in the United Kingdom altogether, while Next have enjoyed buoyant sales. These structural shifts suggest that The Arcadia Group has significantly increased its market share at the expense of M&S and C&A.
4 Richard Hyman, Verdict Market Research, quoted in The Independent, 28/9/99 'Doom and gloom engulfs old high street names' and Guardian, 8/3/99, M&S cranks up retail wars.
5 Abrams and Astill (2001, 1).
6 Tom Peters, CEO, Nike (quoted in Klein, 2000, 240).
7 Called Special Economic Zones in China and maquiladoras in Mexico.
8 See, for example: http://www.tribuneindia.com/2000/20001004/biz.htm
9 Ethical trade initiatives are not simply an altruistic gesture on the part of large corporations; increasingly they make commercial sense too. It has been estimated that the fair-trade market accounted for US$400 million annual retail sales in the late 1990s (Blowfield, 1999).
10 CERES is a multi-stakeholder coalition of unions, investors, pension funds and other public agencies which holds the potential at least to institute change.

References

Abrams, F. and Astill, J. (2001) Story of the blues. *The Guardian*, 29 May.

Anon (1999) Merger mania strengthens retail's power players. *Bobbin*, July 1999, n.p.

Asia Monitor Resource Centre (1998) We in the zone: women workers in Asia's Export processing zones. Asia Monitor Research Centre (*AMRC*), Hong Kong.

Berndt, C. (2001) *El Paso del Norte: Modernization utopias, otherness and management practices in Mexico's Maquiladora Industry.* Unpublished mimeo.

Blowfield, M. (1999) Ethical trade: a review of developments and issues. *Third World Quarterly*, 20, pp. 753–70.

CAFOD (1998) *The Asian garment industry and globalisation.* CAFOD Policy Papers. London.

Catanzarite, L. and Strober, M. (1993) The gender recomposition of the Maquiladora. *Labour Force Industrial Relations*, 32, pp. 133–47.

Coalition for the Charter on the Safe Production of Toys (1996) Our children don't need blood-stained clothes. *Human Rights for Workers Bulletin*, vol 1.2, 23 February.

Cox Edwards, A. (1996) Labour regulations and industrial relations in Indonesia. World Bank Policy Research Working Paper 1640 World Bank, Washington, DC.

Crewe, L. and Davenport, E. (1992) The puppet show: changing buyer-supplier relations in clothing retail. *Transactions of the Institute of British Geographers (IBG)*, 17.

Crewe, L. and Lowe, M. (1995) Gap on the map: towards a geography of consumption and identity. *Environment and Planning A*, 27, pp. 1877–98.

Crewe, L. and Gregson, N. (1998) Tales of the unexpected: exploring car boot sales as alternative spaces of contemporary consumption. *Transactions of the Institute of British Geographers*, 27, pp. 39–53.

Croft, A. (1999) But haven't we seen the stores new look somewhere before? *Evening Standard*, 6 July.

Department for International Development (1998) Minding their own business. *Developments: The International Development Magazine 2* (available online at: http://www.developments.org.uk/data/oz/minding.htm).

Ehrlich, J. (2000) Sweatshop swindlers. *South China Morning Post*, 18 December.

Elson, D. (1984) Nimble fingers and other fables. In Chapkis, W. and Enloe, C. (eds), *Of Common Cloth: Women in the Global Textile Industry*. Transnational Institute.

Entwistle, J. (2000) *The Fashioned Body*. Oxford: Polity.

Fernandez-Kelly, M. P. (1983) *For We are Sold, I and My People: Women and Industry in Mexico's Frontier*. New York: State University of New York.

Figueroa, H. (1997) Blood, sweat and shears. *NACLA Report on the Americas*, vol 29:4, New York, NACLA.

Finch, J. (2001) 'M&S banks on drastic new makeover', *The Guardian*, 30 March.

Fine, B. (1993) Modernity, urbanism and modern consumption: a comment. *Environment and Planning D: Society and Space*, 11, pp. 599–601.

Fine, B. (1995) From political economy to consumption. In Miller, D. (ed.), *Acknowledging Consumption*. London: Routledge, pp. 127–63.

Fine, B. and Leopold, E. (1993) *The World of Consumption*. London: Routledge.

Fuentes, A. and Ehrenrich, B. (1983) *Women in the Global Factory*. Boston: South End Press.

Gereffi, G. (1994) The organisation of buyer-driven global commodity chains. In Gereffi, G. and Korzeniewicz, M. (eds), *Commodity Chains and Global Capitalism*. Westport: Greenwood Press.

Hale, A. (2000) What hope for ethical trade in the globalised garment industry? *Antipode*, 32(4), pp. 349–56.

Hartwick, E. (2000) Towards a geographical politics of consumption. *Environment and Planning A*, 32, pp. 1177–92.

Johns, R. and Vural, L. (2000) Class, geography and the consumerist turn: UNITE and the Stop Sweatshops Campaign. *Environment and Planning A*, 23, 1193–213.

Just-Style (2000) French connection accelerates global brand expansion. *Just-Style*, 14 November.

Kessler, J. (1999) New NAFTA alliances reshape sourcing scene Just-Style.com. Bobbin Publishing Group, 24 Nov 1999.

Kenyon, P. (2000) Gap Nike – no sweat? *Panorama*, 15 October.

Key Note (2000) *Sports Clothing and Footwear Market Report*. London: Key Note.

Klein, N. (2000) *No Logo*. London: HarperCollins.

Leslie, D. and Reimer, S. (1999) Spatializing commodity chains. *Progress in Human Geography*, 23, pp. 401–20.

Lury, C. (1996) *Consumer Culture*. Cambridge: Polity.

National Labor Committee (2000) *Made in China: the role of U.S. companies in denying human and worker rights*. Revised May 2000. Available at: http://www.nlcnet.org/report00/table_of_contents.htm

Ong, A. (1987) *Spirits of Resistance and Capitalist Development: Factory Women in Malaysia*. Albany: State University of New York Press.

Oxborrow, L. (2000) The Marks and Spencer crisis still dominates the textile industry. *Nottingham Evening Post Business Review*, 27 March, p. 7.

Pangestu, M. and Hendytio, M. (1997) Survey responses from women workers in Indonesia's textile, garment and footwear industries. World Bank Policy Research Working Paper 1640. World Bank: Washington.

Phizacklea, A. (1990) *Unpacking the Fashion Industry: Gender, Racism and Class in Production*. London: Routledge.

Raikes, P., Jensen, M. and Ponte, S. (2000) Global commodity chain analysis and the French filiere approach: comparison and critique. *Economy and Society*, 29, pp. 390–417.

Roddick, A. (2000) *Business as Unusual*. London: HarperCollins.

Ross, A. (ed.) (1997) *No Sweat: Fashion, Free Trade and the Rights of Garment Workers*. London: Verso.

Salzinger, L. (2000) Manufacturing sexual subjects: harassment, desire and discipline on a maquiladora shopfloor. *Ethnography*, 1, pp. 67–92.

Smyth, I. and Grijns, M. (1997) Unjuk rasa or conscious protest? Resistance strategies of Indonesian women workers. *Bulletin of Concerned Asian Scholars*, 29, pp. 13–22.

Spielberg, E. (1997) The myth of nimble fingers. In Ross, A. (ed.), *No Sweat: Fashion, Free Trade and the Rights of Garment Workers*. London: Verso, pp. 113–22.

Taylor, C. (2002) Where am I? *The Guardian*, 23 November 2002.

Tjandraningshi, I. (1995) Between factory and home: problems of women workers. In Harris, D. (ed.), *Prisoners of Progress: A Review of the Current Indonesian Labour Situation*. Leiden: Indonesian Documentation and Information Centre, pp. 47–56.

UNITE (2000) *Sweatshops Behind the Swoosh*. Available at: http://www.uniteunion.org/pressbox/nike-report.html

Verdict (1998) *Clothing 1998*. London: Verdict.

Verdict (1999) *Verdict on the High Street*. London: Verdict.

Verdict (2000) *Clothing Discounters*. London: Verdict.

Wark, M. (1997) Fashion as a culture industry. In Ross, A. (ed.), *No Sweat: Fashion, Free Trade and the Rights of Garment Workers*. London: Verso, pp. 227–48.

Williams, C. (1998) The Asian garment industry and globalization, Head of Corporate Communications' Speech, Manchester, 6 May 1998, London: CAFOD.

Wright, M. (1997) Crossing the factory frontier: gender, power and place in the Mexican maquiladora. *Antipode*, 29, pp. 278–302.

Wright, M. (1999) The politics of relocation: gender, nationality and value in a Mexican maquiladora. *Environment and Planning A*, 31, pp. 1601–17.

11 Accounting for ethical trade

Global commodity networks, virtualism and the audit economy

Alex Hughes

Introduction

Part of the promise of ethical trade has rested upon its potential to redress current imbalances in power relations embodied in global commodity chains. Ethical trade seeks to provide a 'third way' solution to inequalities between producers and consumers that have arisen as part of the neo-liberalization of world trade (Barrientos, 2000; Hughes, 2001b). Against the backcloth of a rapidly globalizing and liberalizing trading system, the regulation of commodity chains since the Second World War has fallen increasingly into the hands of the private sector (Watts, 1996; Watts and Goodman, 1997). This so-called period of regulatory crisis (Friedmann, 1993) and private global regulation (Raynolds *et al.*, 1993) has produced widespread and well-known cases of retailers and brand manufacturers in advanced capitalist economies exploiting export producers in economically less developed countries (Appendini, 1999; Cook, 1994; Gereffi, 1994a,b; Hartwick, 1998; Klein, 2000). However campaigns led by consumer groups and non-governmental organisations (NGOs) most recently have sought to challenge this corporate-led regulatory (dis)order. Notable examples include the Stop Sweatshops Campaign (Johns and Vural, 2000) and Christian Aid's exposure of the poor treatment of 'Third World' food producers by manipulative supermarket chains (Orton and Madden, 1996). Supported by increasing media exposure of exploitative trade, campaigns have paved the way for new ethical trading developments which aim to ensure that relationships between 'North' and 'South' neither harm worker welfare nor damage the environment at sites of export production. Ethical trading initiatives contrast markedly with NGO-led fair trade projects (such as those of Oxfam and the Fair Trade Foundation) which attempt much more profoundly to assist producers in seizing greater control of the supply chain (Renard, 1999; Whatmore and Thorne, 1997). Less a radical challenge to the organization of power in the global commodity chain, and more a corporate knee-jerk reaction to the adverse effects of consumer campaigns on profitability, ethical trading initiatives aim simply to ensure *minimum standards* in environmental protection and labour rights within an existing model of trade liberalization and private sector dominance.

The aim of this chapter is critically to review the role of ethical trade in refiguring relations of power in the global commodity chain. Discussions within the existing literature on ethical trade (Barrientos, 2000; Barrientos *et al.*, 2000; Blowfield, 1999; Hale, 2000; Hale and Shaw, 2001) have tended to focus upon the driving forces behind the emergence of ethical trade, reviews of existing initiatives and the development of best practice. While these themes are of undoubted importance in the practical quest to improve conditions of export production, it is timely to take stock of the role of ethical trade in re-regulating global commodity chains and to broaden debate about the movement's politics. The chapter therefore explores some of the regulatory contradictions and complexities inherent in ethical trading programmes, focusing in particular upon the creation of a new 'audit economy'. Because the credibility of ethical trading initiatives depends critically upon the monitoring and verification of environmental and labour standards at sites of production (Zadek, 1998), social, environmental and ethical auditing has become an increasingly important part of the process. At the same time, the use of corporate-led auditing practices to guarantee core labour and environmental standards has attracted criticism for its marginalization of traditional government legislation (Klein, 2000). Commentators have highlighted the practical challenges of monitoring ethical trade, exemplified by the problems even of identifying producers to audit amidst dense webs of subcontracting supply relations (Hale, 2000; Hale and Shaw, 2001). This chapter, however, seeks to provide a broader critique of the politics of the audit economy and its application to the ethical trading project. Elsewhere, I have applied notions of governmentality to the critical study of ethical trade and auditing (Hughes, 2001a) – here I turn my attention to an alternative set of concepts.

Miller (1998, 2000) has argued that the turn of the twenty-first century witnessed the development of a new political economy built on 'virtualism', whereby the old capitalist order began to give way to an economy shaped increasingly by powerful, abstract models. The direct influence of consumers upon the organization of the economy is being replaced by corporate strategies executed *in the name of* the virtual consumer rather than in response to 'real' consumer preferences, practices and desires. With its focus on abstract knowledge (such as that crystallized in ethical codes of conduct) and its re-conceptualization of consumer power in the commodity chain, this theoretical framework offers a potentially productive means of conceptualizing the auditing of ethical trade.

The chapter draws upon recent research into an ethical trading programme in the Kenyan cut flower industry; an industry that has become a major supplier to European markets over the past thirty to forty years (Hughes, 2000, 2001a). I suggest that theories of virtualism are useful in alerting us to the regulatory complexities and contradictions wrapped up in auditing the work of the Kenya Flower Council – a national-level, business-led organization that is designed to construct, monitor and promote environmental and labour standards in the country's cut flower export industry.[1] Before presenting a discussion of auditing in this local context, and the ways in which it might represent a 'virtual' version of ethical trade, I first outline Miller's theory of virtualism and its potential applicability to ethical auditing more broadly.

Virtualism and the audit economy

Miller (2000, 200) argues that we are witnessing 'a new chapter in the history of political economy'. He suggests in particular that the unprecedented and over-whelming rise in the power of abstract, economic models to shape the workings of the economy currently requires explanation (see also Carrier, 1998). The term 'capitalism' has been used for so long in the labelling of so many different eco-nomic forms in the world that Miller (1998, 2000) seeks a theory more specific to the present character of the economy. Following a mode of investigation used in earlier work (Miller, 1987), Miller turns to Marx and the dialectical approach for some answers. While acknowledging the limitations of grand narratives, he nonetheless searches for an overarching explanation for the economic changes that constitute the 'dominant discourse' of today's political economy (Miller, 2000, 210).

Perhaps controversially, Miller (2000) proposes that consumption can be seen as the *negation* of some of the homogenizing and alienating forces of early cap-italism noted by Marx in the nineteenth century. In more recent transformations, the economy increasingly has been influenced by forces of abstract modelling:

> [I]f it is the case that consumption and other forces have been used to counter the kinds of capitalism that Marx described, then I suspect Marx himself would be on the search for what might be called the negation of the negation. There should be some new force emergent that is based on the contradictions of these earlier forces – perhaps the contradictions of con-sumption. This would be some newly risen form of abstraction that will come to replace capitalism but must necessarily be more extreme than capi-talism, more abstract and threaten to be still more dehumanizing until it in turn can be negated. Marx today would surely not be turning back to capi-talism as its last predecessor but looking forward to this new target in its emergent form.
>
> (Miller, 2000, 200)

This 'new target' is a contemporary era of virtualism that has both absorbed and replaced capitalism. Miller (2000, 197) suggests that two basic principles govern the 'fundamental attributes of contemporary virtualism'. The first concerns the power of abstract, economic models to the extent that '…models which are thought to be descriptive of economic relations have become so powerful that they become in and of themselves the forces that determine economic relations' (197). The second relates to the prevalence of the 'virtual consumer' over 'real', embodied consumption practices and consumer preferences, at least in terms of the ways in which economic action in the productive and distributive spheres is justified and rationalized. In order to demonstrate the grounding of these practices in contemporary economy and society, Miller (1998, 2000) reviews five different case studies, including structural adjustment and the auditing of universities. The application of the theory of virtualism to these two cases in particular provides some clues as to how principles of virtualism might be shaping the auditing of ethical trade.

The example of structural adjustment is perhaps the most significant and most powerful case study. The strategies and practices of key global institutions such as the IMF, the World Bank and the World Trade Organisation are based firmly on economic models of neo-liberalism and market efficiency as a means of guiding economic development. Furthermore, these models are executed and justified in the name of the consumer:

> The whole point of the market is that it is argued to be the sole process which might bring the best goods at the lowest prices to the consumer who thereby is the ultimate beneficiary from this process…[But] the consumer of economic theory is not an actual flesh and blood consumer. Indeed, as practitioners of neo-classical economic theory constantly remind us, they make no claim to represent such flesh and blood consumers. They will tell us that their consumers are merely aggregate figures used in modelling.
>
> (2000, 203)

A second example of virtualism is provided by the sphere of auditing, with a focus on the auditing of universities. Drawing on Strathern (1997), Miller (2000) reflects upon the current dominance of research and teaching quality audits in the UK higher education system. Beyond the fields of economics and finance, virtualism prevails in the use of quality audits in this part of the public sector. Typically, teaching and research quality audits (including their rapidly expanding paper trails) are said to be driven by the needs of students and tax-payers as 'consumers'. However, once more the consumer is virtual:

> all of these auditing procedures are justified on grounds that lead to the con-sumer, either as the taxpayer getting value for money or as the actual recip-ient of the services…Yet as any academic knows from experience the main effect on students is a loss of teaching time, since they are replaced by those who stand for them in aggregate form, that is the managers who carry out audits.
>
> (Miller, 2000, 207)

Not only does the role of the virtual consumer feature prominently in the governance of higher education, but so too do forces of abstraction. Audits seek to achieve the 'control of control', using practices of checking, monitoring and collecting evidence developed from the field of financial accounting (Miller, 2000). The process of abstraction is visible in higher education through the increasing confinement of university activity into measurable units. As Strathern (1997, quoted in Miller, 2000, 207) notes, 'measures become reduced to targets' (308), 'descriptions become prescriptions' (312) and 'any sense of contradiction or ambiguity is replaced by canons of clarity and itemization' (315). Measures of academic work therefore begin to drive, as well as describe, workplace activity.

Such arguments are echoed in broader observations made by O'Neill (2002). She suggests that a 'crisis of trust' has led to the emergence of a 'new accountability culture', which now controls and regulates both the public and private sectors. However, O'Neill (2002) strongly criticizes the current methods used for accountability, with their overemphasis on total transparency, standardization and performance indicators. Like Strathern and Miller, O'Neill argues that the onerous task of producing detailed paperwork frequently diverts the attention of public sector workers away from actually improving real standards in an appropriate way. Moreover, detailed paperwork often conforms to contradictory targets and inappropriate measures of performance, rather than helpful and 'intelligent' forms of accountability (O'Neill, 2002). The arena of ethical trade is certainly not immune to these complexities and contradictions of the 'new accountability culture'.

The theory of virtualism, with its emphasis on the simultaneous power of abstraction and the virtual consumer, would seem to offer potentially productive ways of thinking about the auditing of ethical trade. In the following sections of the chapter, I reflect upon notions of virtualism in the light of ethical trade practices in the Kenyan cut flower industry in order to provide an insight into the politics of ethical trade. I begin with a brief overview of this particular globalized commodity network, then trace through the role of the virtual consumer in the network, and finally unravel processes of abstraction epitomized by the Kenya Flower Council's code of practice.

Virtually ethical?: the auditing of the Kenyan cut flower industry

Growing networks: the emergence of the auditor in the cut flower trade

The production of flowers for European consumers historically has been located in the Netherlands. However since the 1960s the production of commercially grown cut flowers for European and North American markets has shifted to economically less developed countries such as Colombia, Ecuador, the Gambia, India, Kenya, Uganda and Zimbabwe (Coulson, Chapter 7, this volume; Maharaj and Dorren, 1995). This extending spatial reach of commercial cut flower production gathered pace during the 1980s, due in part to Structural Adjustment Programmes and pressure on developing countries to produce high-value export crops in order to generate foreign exchange (Barrett *et al.*, 1999; Jaffee, 1994; Maharaj and Dorren, 1995; Watts, 1996).

Kenyan flower production began in the 1960s, initiated by European settlers on semi-marginal land (Detmers and Kortlandt, 1996). Since then the industry has grown rapidly, largely in an attempt to cover the country's US$250 million trade deficit affected by a drop in prices for traditional export commodities of tea and coffee (Maharaj and Dorren, 1995). Expansion continued rapidly throughout the 1990s. In 1992, the value of Kenyan cut flower exports was US$61.48 million

(Maharaj and Dorren, 1995). By 1998, this figure had risen to US$80.9 million.[2] In volume terms, Kenyan cut flower exports amounted to 30,229,033 kg in 1998, with 65 per cent of this product going to Holland,[3] 16 per cent to the United Kingdom and 9 per cent to Germany (Kenya Flower Council, 1999). Among the most common types of flowers grown are roses, statice, alstroemeria and carnations, which predominantly are cultivated in the main growing regions of Lake Naivasha and Thika. Over 800 hectares of Kenyan land is devoted to the cultivation of cut flowers, with production dominated by three large farms, two of which occupy around 250 hectares of land each. Remaining production takes place in thirty medium-sized farms of varying sizes above 5 hectares and several hundred smallholdings of less than 5 hectares each. The three largest growers are situated in Naivasha, drawing on the resources of Lake Naivasha and attracting thousands of migrant workers from all over Kenya in search of employment. The two largest producers, Oserian and Sulmac, each employ over 5,000 workers, a small proportion of which constitute occasional rather than permanent labour. It is these large growers that are most able financially and organizationally to meet the strict demands of UK supermarket chains (Barrett *et al.*, 1999; Hughes, 2000, 2001a).

For farms supplying UK markets, the interests of large retail chains (Sainsbury, Tesco, Safeway, Asda, Waitrose and Marks & Spencer) increasingly have influenced the ways in which they do business. Growers have been placed under considerable pressure to conform to retailers' requirements for traceability in the supply chain, in part as a result of the 1990 Food Safety Act (Barrett *et al.*, 1999). Moreover, growers also have been forced to produce flowers that conform to the ever-changing, design-based requirements of retailers. Such demands can be extremely difficult to meet amidst climatic constraints, the time taken to breed new flower varieties, and heightened international competition from a growing number of producers (Hughes, 2000). This pressure on growers is further intensified by the seasonality of cut flower consumption, linked to festivals such as Valentine's Day, Mother's Day, Easter and Christmas. Seasonality is a major force behind the employment of large numbers of temporary workers at times of peak demand, along with pressure on employees to work overtime (Appendini, 1999; Meier, 1999). UK retailers increasingly have manipulated the complex supply networks of the global cut flower trade, placing ever greater stress on the production process in order to develop markets (Hughes, 2000).

At a time when the international production of commercially grown flowers had truly become 'big business' (Maharaj and Dorren, 1995), the mid-1990s witnessed a spate of media-generated public concern about working conditions on flower farms in developing countries. Articles in the European press – and particularly in the United Kingdom – highlighted economic uncertainty, low wage levels for workers, cases of child labour, and risks to both the health and safety of workers and the environment due to pesticide use on farms (Bolger, 1997; Durham, 1996; Wolf, 1997).[4] Journalists attempted to foster public concern by contrasting conditions at sites of production with the symbolic meanings of luxury, design and gift-giving associated with this high-value commodity at sites of consumption (Shakespeare, 1995; Stewart, 1994). Such media attention sent shock waves through the global flower industry. Combined with related pressure from NGOs to improve working conditions, this situation triggered the establishment of an organization set up to promote more ethical production in the industry: the Kenya Flower Council.

In 1999, the Kenya Flower Council had twenty-two member farms and was led by an independent Executive Officer in Nairobi, with help from a small number of assistants. The organization had eight Council Directors drawn from the largest of the farms. Most members continue to be the large farms located in Naivasha and Thika, though others are situated in Kericho and Nyahururu. Significantly, there are also six Associate Members comprising key overseas buyers of Kenyan flowers in Holland, Germany, Switzerland and the United Kingdom.

Once the Kenya Flower Council was established, its key members drew up a code of practice. Codes of practice (or codes of conduct) are the most common starting point for ethical trading initiatives. Put simply, they embody the minimum standards of environmental protection and worker welfare for industry to follow, though code content varies considerably between different ethical trading organizations (Blowfield, 1999). While organizations like the UK's Ethical Trading Initiative (Barrientos, 2000; Barrientos *et al.*, 2000) base their codes solely on labour standards, others such as the Kenya Flower Council also incorporate environmental clauses (Kenya Flower Council, 1998). I address the Code of Practice in further detail below, but for the moment it is important to note that in order to receive credibility from retail buyers and consumers in the commodity chain, grower compliance with the code requires monitoring. While monitoring is conducted to some extent through internal audits by the Kenya Flower Council Executive Officer with the help of an assistant, the internal audit is itself checked and verified by an internationally renowned, specialist auditing firm, BVQI. The emergence of the Kenya Flower Council therefore represents the entrance of a new set of agents – the auditors – into the global commodity chain for cut flowers.

The dynamics of the commercial flower trade are strongly shaped by the associated activities of design (flower arranging) and flower breeding. The range of agents involved in the trade thus produces a complex network of multi-stranded exchange relationships (Hughes, 2000). The recent involvement of social and environmental auditors (both 'internal' and 'external') in the cut flower industry adds further complexity to this network. Rather than conceptualizing the cut flower industry as a linear global commodity chain, it is more appropriate to use the term 'global commodity network'. The chapter now turns to consider the ways in which the work of the Kenya Flower Council in this network is driven first by the virtual consumer and second by forces of abstraction epitomized by both the Code of Practice and rationalized audit inspections.

Auditing in the name of the virtual consumer

Practices of auditing have infiltrated 'almost every branch of governance' (Power, 1997, quoted in Miller 2000, 207). The monitoring of codes of conduct in global supply chains is no exception to this trend. As Anita Roddick, founder of The Body Shop, proclaims: 'Ten years ago, the idea of an environmental audit was popularly regarded as the provenance of the lunatic fringe. Now the principle has been internationally accepted, largely due to the power of public opinion' (2000, 69).

However the extent to which 'the power of public opinion' actually drives ethical trading organizations such as the Kenya Flower Council is debatable. It is generally argued that calls for greater accountability lie behind demands for the auditing of ethical trade; accountability of retailers to stakeholders, including the public. This suggests that consumers require *proof* that companies are sourcing from overseas suppliers who enforce minimum standards in worker welfare and environmental protection. Pentland (2000, 307) argues that 'we are experiencing...a movement along a continuum from a society that trusts everything and audits nothing towards a society that trusts nothing and audits everything'. For ethical trade, this implies that audits can be used by companies to assure potentially disbelieving consumers that ethical codes of conduct are applied in global supply chains.

For Hartwick (2000, 1178) audits represent the kind of 'geographical detective work' that promises to reconnect distantiated producers and consumers in the commodity chain. Implicit in Hartwick's (2000) argument is the assumption that consumers themselves should conduct this detective work. However, research into ethical trading initiatives reveals that it is far more common for supply chain auditing to be led by the private sector. Thus the extent to which politically progressive links are actually made in the global commodity network for cut flowers is limited. Drawing on Miller's (1998, 2000) theory of virtualism, I suggest that it is the media, campaigning organizations and UK supermarket chains, acting as virtual consumers, who actually drive the work of the Kenya Flower Council and its auditing practices.

The forces behind the development of ethical trading initiatives are widely regarded as emanating from the realm of consumption (Crewe, Chapter 10, this volume; Klein, 2000; MacKenzie, 1998). UK market research has suggested that increasing numbers of consumers are now concerned about ethical issues (Freidberg, 2003). While there is clearly a need for more research in this area, the Kenya Flower Council example indicates that it is rarely consumers themselves who directly demand codes of conduct and auditing. Rather, negative media exposure and NGO pressure, together with demands from retail buyers dictate the need for ethical strategies. Recent media attention paid to unethical practices in global supply chains is acknowledged as a major force motivating corporate monitoring, auditing and reporting (Klein, 2000). Exposure of incidences of child labour in the global supply chains for footwear sold through Marks & Spencer is a case in point (O'Riordan, 2000). As I have indicated, radio documentaries and articles in the European press adversely affected the public image of Kenyan flower producers during the mid-1990s. Managers of the commercial farms acknowledged that the creation of the Kenya Flower Council was a direct response:

There were maybe six farms who set up the Kenya Flower Council three years ago. We got together and we decided that, because there was no established association with a code in the country, that we were at risk to outside press, various people, er, pressure groups, to come and criticise what we do because there wasn't any cohesive organisation. So the six of us got together and discussed what we should do. So we decided to set up the Kenya Flower Council.

The Kenya Flower Council's Code of Practice, drawn up by the directors of member farms, directly responds to issues addressed in the press articles and by pressure groups. Concern about worker health and safety, along with the effects of commercial production on Lake Naivasha as a result of pesticide use is reflected in the seven key headings of the code: 'farm management responsibilities and proper documentation', 'crop protection strategy', 'safe use of pesticides', 'application of pesticides and protection of workers', 'transportation, storage and disposal of pesticides', 'general worker welfare', and 'protection of the natural environment'. Within the detailed clauses there also is an attempt made to respond to the increasingly stringent demands of the UK supermarket chains with respect to documentation, traceability and auditing:

The background [to the code] was very much drawn particularly from the UK supermarket audit procedures...They're very much geared into the same sort of things – worker safety, how you're looking after your chemicals, how you're looking after your workers...And seeing as we sell probably 70% of our produce in the UK, then it's fairly obvious that that was the major influence.

The combined forces of the press, campaigners and supermarket buyers can therefore be seen to influence the Kenya Flower Council's ethical trading efforts. Furthermore, Kenyan growers continually are required by retailers to respond to the latest round of media coverage, some of which now addresses the effectiveness of the new codes and auditing systems. In a recent article, Duncan Green from the NGO, CAFOD, describes how:

...managers pore over the latest suppliers' briefing from Marks and Spencer – an annotated copy of a recent highly critical *Guardian* article about the flower industry in Colombia. Scribbled marks in the margin say 'answer accusations with evidence – don't get defensive.'

(2001, 25)

While the motivation for constructing the Code of Practice and the system for ethical auditing is therefore quite rightly suggested to derive from the sphere of consumption, the consumers of cut flowers themselves are less directly involved. Rather, it is the media, NGOs acting *on behalf of* the consumers and the supermarket chains who call for developments in ethical trade. Moreover, the 'Environment Friendly' logo, which Kenya Flower Council members are permitted to use following the successful audit of their production sites, appears only on business letters and the boxes used for flowers in transportation and distribution. At the time of writing, the logo does not appear on the cellophane wrapping around the individual bouquets sold to consumers (see also Coulson, Chapter 7, this volume). While country of origin information sometimes appears on retail products, the ethical status of the producer goes unremarked. Consumers themselves therefore appear to be written out of the 'circuit of knowledge' (Cook *et al.*, 2000) shaping the ethical trading of cut flowers. One of the buyers of Kenyan cut flowers, representing a major UK supermarket chain, appears to support this view:

> I'm not sure the pressure is ever from our customers...We don't get thousands and thousands of customer queries about how you're treating your workers at source. [...] We're not trying to have it as a, it's not a sales pitch saying that we've got ethical trading. It's basically an insurance policy.[5] I think it's just to satisfy sales that if anything does happen, if there are any issues, we can say very confidently that we have a policy...I think it's probably being led by other supermarkets...[...] if an issue, a food safety issue or a worker welfare issue blows up in the press, then the quicker we can react, and the more positively we can react, then we can reduce the damage that potentially could be done to our company in terms of sales, turnover, you know. So it's almost not something we want to stand on the parapet and shout about. We don't need to do that – our customers aren't interested. They're only interested when it goes wrong, and that's all protocols and codes of practice are about really.

Further research is needed to investigate the ethical knowledges of consumers. However in light of the Kenyan study, evidence suggests that as in the case of retail mergers and category management (Cook *et al.*, 2000; Miller, 2000), it is the supermarket chains, influenced by media and NGOs, who appear to represent the virtual consumer forcing the development of codes and auditing. In the case of the cut flower commodity network, then, there are limits to the extent to which codes and auditing can redress unequal power relations between producers, retailers and consumers. Practices of ethical auditing are still very much conducted as part of a 'private interest model of regulation' (Marsden and Wrigley, 1995, 1996; Marsden *et al.*, 2000) shaped by the virtual consumer.

The code, the audit and abstraction

With reference to auditing more broadly, Miller (2000, 207) notes that '...the audit explosion also conforms to the first characteristic of virtualism which is that it gives power and authority to abstract modelling'. The power of abstraction to shape the economy and society is not only visible in the economic models of structural adjustment, but also is apparent in new models of governance based on auditing. As Carrier (1998, 25) suggests:

> Virtualism is the attempt to make the world conform to an abstract model, so abstraction is virtualism's foundation. But it is important not to restrict attention only to systems of abstract thought like neo-classical economics. Such systems are important, but so is practical abstraction, abstraction in daily life and practice.

In this section of the chapter, I reflect upon the process of abstraction at work in the Kenya Flower Council's Code of Practice. The third edition of the Code used at the time of my study was a hefty 34-page document consisting of numerous detailed clauses under the seven key headings introduced earlier. The broad coverage of the code, incorporating specific standards for both worker welfare and environmental protection, therefore contrasts with the Base Code of the UK's Ethical Trading Initiative with its exclusive social focus on nine core ILO conventions (Barrientos *et al.*, 2000; Hale, 2000). But as with all codes of conduct, it necessarily embodies a set of abstract rules and conventions for business to follow. The initial construction of the code can be viewed as a process of abstraction, whereby directors of the leading commercial flower farms strategically determine what the clauses of the code should be. Clauses in the code thus represent the programmatic aims and ideals of responsible flower growing. Conforming to Miller's (2000) theory of virtualism, this code is both a model designed to reflect the practical business of flower growing and yet also is aimed at changing the way in which this business is performed. Procedures for auditing farms are then set up in order to make the code practicable. In other words, auditing is the key cog in the system for developing commercial growing in line with the Code of Practice.

Each of the Kenya Flower Council's members is subject to an audit by the Executive Officer of the organization. Once a farm's initial application for membership is accepted, they are required to pay a small fee. At the same time, they are handed a copy of the Code of Practice and given one year in which to comply with its requirements. Within that year, they have to be audited. The Executive Officer and assistant conduct an internal audit, touring the farm equipped with eight pages of questions attached to a clipboard, which mainly elicit results in the form of nominal and ordinal data corresponding to the code's clauses. The Officer returns to the Flower Council's head office and writes a report for the farm based on these results. If there are weaknesses, the farm is asked to make the requisite improvements and is subject to a repeat visit within a few weeks. Once the farm is deemed to have passed, they are given the Silver Award (a new Gold Award has been developed since the time of my study). The member is then due for a re-audit every six months. Audits themselves take half a day to a day, involving extensive preparation on the part of the farms. Crang (2000, 214) has touched upon the 'textual and interpersonal stagings of organisational achievement' associated with quality audits and assessments. My observation of a farm re-audit confirmed that such stagings are very much a part of the Flower Council audits, as numerous pieces of paperwork and files are laid out for close inspection, and key members of the workforce are interviewed. While the highly performative acts of the audit must be acknowledged, the crystallization of the code's key clauses into a checklist and set of simple questions illustrates the process of abstraction once again, whereby complex organizational activities are frozen into simple, predetermined categories of data.

Power (1997) has suggested that procedures of audit often become separated from the practical realities of the worlds being audited. This 'decoupling' occurs 'when an audit process is disconnected from what is really going on' (Power, 1997, 96). As such, decoupling might represent the process by which codes and audits are abstracted from real, lived experiences in the workplace. In the remainder of this section, I illustrate how decoupling appears to be at work in the Kenyan cut flower industry and the application of the Code of Practice. Limits of space make it impossible to review all elements of the code and all aspects of auditing, thus my focus is upon a single key section of the code: 'general worker welfare'.

Twenty-four clauses appear under the heading of 'General Worker Welfare'. Here, my particular interest is in provisions concerning housing occupied by flower farm workers. The following two clauses appear under section 6 of the Code of Practice:

> All employees are entitled to, at least, the minimum wage, fixed annually by the Government Regulation of Wages (Agricultural Industry) order and/or by those entities representing the worker's interest. Wages should be paid in cash at the agreed time and in full. Workers should be given a detailed pay slip.[6]
>
> Every employer shall at all times, at his (sic)[7] own expense provide reasonable housing accommodation with adequate running water and toilet facilities for each of his employees or shall pay to the employee such sufficient sum as rent allowance in addition to his wages or salary as will enable the employee to obtain reasonable accommodation.[8]
>
> (Kenya Flower Council, 1998, 10–11)

Farm audits conducted by both the Kenya Flower Council and their external auditor, BVQI, include checks for compliance with both of these labour conventions. However, at the time of my study checks were entirely verbal. Management were asked if they complied with the requirements, requiring a yes/no response, and a small sample of workers were interviewed briefly about working and living conditions. Interviews were brief and the process visibly intimidated workers, who revealed few of their experiences to the auditor. Although current research conducted by ethical trading organizations is seeking to improve methodologies for worker interviews in social audits (Burgess and Burns, 1999), my discussions with workers suggested that many specific issues of housing and living conditions remained untouched by the code and were hidden from view in the audit process. Workers quite simply '[fell] out of account' (Power, 1997, 100). Code clauses regarding wages and housing dictate minimum standards that fall far short of workers' everyday needs.

Clauses often are abstracted from the 'social and practical contexts' of workers (Carrier, 1998). The two largest commercial farms, Oserian and Sulmac, along with one or two other companies, house the majority of their workers on the site of the farm, adopting a paternalistic strategy towards worker welfare and business ethics (Crossley, 1999). Most other Kenya Flower Council member farms elect to pay workers a housing allowance, normally about 600 Kenyan shillings per month. In both situations the standard of housing is poor. As an Assistant Supervisor explained: 'A house with electricity ranges from 800 shillings and above [per month], but the maximum house allowance I get here is 608 shillings, so I have to get some money from my pocket in order to pay the rent, in order to stay in the house whereby you can have enough water and electricity.'

There are also problems for workers housed on the farm site, involving often cramped living conditions. A flower packer explained that, 'The houses are still being built, but there are a lot of workers, which means the workers are outnumbering the number of houses' (interview, 15 September 1999). For a supervisor, conditions were deemed to be adequate. She was able to share a house with her husband, who also works on the farm as a crop sprayer, and their small child. She said, 'well, of course, we would like it much bigger, but for now it is just okay' (interview, 15 September 1999). However, for general workers working and living on the farm away from their families, and forced to share accommodation with other workers, the problem of living space appears more acute. A woman, who performs general work in the carnation fields, elaborates: 'Supposing that maybe you are married and you are given a house to two or three of you and then maybe your husband comes to visit you, it's just a problem.'

Living in close proximity to production also is problematic: 'Where our premises are, there are greenhouses...The way I see it is that, it's not a very, very good living place. Sometimes that smell from the spraying...Maybe they are spraying just about ten metres from my house.'

Another problem associated with housing and living conditions concerns availability of clean water. Although this issue is touched upon in the Code of Practice, it is not pursued in an extensive way in farm audits. Off-site locations often have an inadequate provisioning system for water:

> In Karagita, the big problem with houses is they don't have tap water. Therefore, we have to buy water daily and...I have seven children, and for water to cater for that family, I need a lot of water to buy. And the house allowance I am given is not enough to cater for the water and the house. And the sanitation there is a problem because most of the houses there are very dusty. There is a lot of dust there and then I have to go to the house and wash daily because of the chemicals I am using at work.
>
> (interview with flower farm supervisor, 7 September 1999)

A final issue for workers living off-site relates to the location of their housing in Karagita. A few miles further away in the nearby town of Naivasha, the standard of housing is deemed to be higher and the costs slightly lower. However, transport is provided by the company only as far as Karagita, and the housing allowance plus wages cannot stretch to cover the fares on the matatu (local public transport) to Naivasha Town (interviews, 7 September 1999). The housing and basic living circumstances of workers are affected by a wide range of issues, including household composition, transport requirements and the provision of utilities. It was clear from my discussions with workers that the clauses in the Code of Practice do not take account of the social and practical contexts of workers' everyday lives. They represent minimum standards, and epitomize what Miller (2000) has discussed as 'disembedded knowledge'. Given that the code and the audits represent the key means by which the global commodity network for cut flowers is regulated (Hughes, 2001a), minimum standards become central benchmarks against which labour standards in the industry are measured. In this respect, the code and the audit appear mournfully disconnected and 'decoupled' from everyday contexts. Falling far short of sufficiently improving standards of living for workers, sections in the Code of Practice concerning wages and housing represent some of the 'dehumanising forces' of abstraction associated with the contemporary moment of virtualism (Miller, 1998, 2000).

Conclusions

This chapter has sought to extend study of ethical trade beyond policy-orientated concerns with best practice to a critique of its politics. While the establishment of the Kenya Flower Council has led to some practical alterations to worker health and safety and environmental protection measures (Hughes, 2001a), the chapter also has emphasized the complexities and contradictions inherent in ethical strategies. I have argued that while the ethical trading strategies of the Kenya Flower Council are deemed to be driven by public concern and consumer demands, they are in fact directed more strongly by retailers' concerns about profitability and the effects of negative media exposure on their sales figures. Further, I have indicated the abstraction of the Code of Practice's clauses from the lived experiences of workers.

The regulatory complexities embodied in the work of the Kenya Flower Council arise out of a much broader political-economic context than hitherto appreciated in the ethical trading literature. While the full theorization of this neo-liberal context is a longer and more complex task than can be accomplished in this chapter, I have suggested that the theory of virtualism (Miller, 1998, 2000) provides a useful conceptual tool for understanding the driving forces behind ethical trading developments. As I have argued, the audit economy that makes the ethical trading movement practicable is shot through with forces of virtualism, ranging from the supermarkets acting as virtual consumers, to the disembedded knowledge characteristic of the Code of Practice. There are limits to the explanatory reach of the theory of virtualism, not least associated with the teleological tendencies of its dialectical approach and the resulting marginalization of political agency and resistance. Indeed, the polemical nature of Miller's (1998, 2000) arguments must be borne in mind. However, the notion of virtualism at least alerts us to some limitations of ethical trade as it is currently being developed. While theories of governmentality perhaps say more about the politics and *practices* of ethical trade (see Hughes, 2001a), and while there is a need for further theorization of the ways in which ethical trade might be more radically and progressively practised to serve the interests of workers in the commodity chain, ideas of virtualism allow for the situation of ethical trade within a wider political-economic context.

Applying notions of virtualism to the case of ethical trade also can help to gain critical purchase upon power relations in the global commodity network. As we have seen, the emergence of auditors and auditing in the cut flower trading system adds another set of agents as well as further organizational complexity to the network, while internal and external auditors have become the main actors regulating the sector. The concept of virtualism alerts us to the ways in which retailers, in the guise of the consumer, drive a re-regulation of the commodity network and thereby deny the possibility of more equitable connections between producers and consumers. Further, it opens up an understanding of how processes of abstraction, bound up in conventions of codes and audits, can at the same time act to deny producers and workers the level of empowerment promised by ethical trading intention and rhetoric. Reducing worker welfare to the most simple, convenient and measurable targets is not the most effective way to improve labour standards (cf. O'Neill, 2002). While piecemeal improvements are being made in the areas of worker health and safety and the reduction of pesticide use (Hughes, 2001a), current ethical trading practice cannot yet be heralded as the answer to redressing large-scale economic and social inequalities experienced in the workings of global commodity networks. With its operation remaining firmly within the neo-liberal model of private interest regulation, these new trading initiatives are best labelled as 'virtually ethical'.

Acknowledgements

My thanks to Suzanne Reimer and Nick Henry for comments on an earlier draft of this chapter. I am grateful to The Nuffield Foundation for funding research on the Kenyan cut flower industry (Award No. SGS/LB/0270) and to Joseph Makau for providing translation in field interviews. Finally, I would also like to acknowledge the cooperation of the Kenya Flower Council throughout the research.

Notes

1 Field research was sponsored by The Nuffield Foundation and took place in 1999, incorporating interviews both with the managers of 13 of the 22 commercial flower farms which belonged to the Kenya Flower Council at the time of the study and with a cross-section of employees in five member farms. Discussions also were held with other key actors, including the Executive Officer of the Kenya Flower Council, BVQI (the international auditing company responsible for monitoring the industry in Kenya), the Kenya Plantation and Agricultural Workers Union, the Lake Naivasha Riperian Association, and buyers representing four of the leading UK supermarket chains who purchase produce from these Kenyan farms. I also attended and observed a farm audit by the Kenya Flower Council in Nyahururu.

2 The figure is based on Kenya Flower Council calculations. In 1997 exports stood at US$81.7 million, but the Kenyan industry suffered a drop in productivity during the following year as a result of the EL Nino (Kenya Flower Council, 1999).
3 From Holland, many of these flowers move through the Dutch auctions and on to other final destinations.
4 Interviews with Kenyan growers revealed that similar press reports also appeared in Germany.
5 This point resonates with O'Neill's (2002) observation that much auditing is defensive in nature.
6 This basic minimum wage at the time of my study was 1,716 Kenyan shillings per month – the equivalent of just under £15.50. However, Kenya Flower Council members pay two to three times this wage to their general workers (interview with Executive Officer, 2 September 1999).
7 The majority of farm owners and managers are male in contrast to the majority of farm workers who are female.
8 The minimum housing allowance for workers at the time of my study was 500 Kenyan shillings (£4.50) per month. In Karagita, where the majority of workers in Naivasha live, the cheapest and smallest accommodation costs 250 Kenyan shillings per month, but houses with electricity cost upwards of 800 Kenyan shillings per month.

References

Appendini, K. (1999) 'From where have all the flowers come?' Women workers in Mexico's non traditional markets. In Barndt, D. (ed.), *Women Working in the NAFTA Food Chain: Women, Food and Globalization.* Toronto: Second Story Press.

Barrett, H., Ilbery, B. W., Browne, A. W. and Binns, T. (1999) Globalization and the changing networks of food supply: the importation of fresh horticultural produce from Kenya into the UK. *Transactions of the Institute of British Geographers*, 24, pp. 159–74.

Barrientos, S. (2000) Globalisation and ethical trade: assessing implications for development. *Journal of International Development*, 12, pp. 559–70.

Barrientos, S., McClenaghan, S. and Orton, L. (2000) Ethical trade and South African Deciduous fruit exports – addressing gender sensitivity. *European Journal of Development Research*, 12, pp. 140–58.

Blowfield, M. (1999) Ethical trade: a review of developments and issues. *Third World Quarterly*, 20, pp. 753–70.

Bolger, A. (1997) Unions call for code to protect flower workers. *Financial Times*, 9 May, p. 4.

Burgess, P. and Burns, M. (1999) *ETI Pilot Interim Report.* London: Ethical Trading Initiative.

Carrier, J. (1998) Abstraction in western economic practice. In Carrier, J. G. and Miller, D. (eds), *Virtualism: a New Political Economy.* Oxford and New York: Berg, chapter 1, pp. 25–47.

Cook, I. (1994) New fruits and vanity: symbolic production in the global food economy. In Bonanno, A., Busch, L., Friedland, W. H., Gouveia, L. and Mingione, E. (eds), *From Columbus to ConAgra: The Globalization of Agriculture and Food.* Kansas: University Press of Kansas.

Cook, I., Crang, P. and Thorpe, M. (2000) Have you got the customer's permission?: category management and circuits of knowledge in the UK food business.

In Bryson, J. R., Daniels, P. W., Henry, N. and Pollard, J. (eds), *Knowledge, Space, Economy*. London and New York: Routledge.

Crang, P. (2000) Organisational geographies: surveillance, display and the spaces of power in business organisation. In Sharp, J. P., Philo, C., Routledge, P. and Paddison, R. (eds), *Entanglements of Power: Geographies of Domination/Resistance*. London: Routledge.

Crossley, D. (1999) Paternalism and corporate responsibility. *Journal of Business Ethics*, 21, pp. 291–302.

Detmers, M. and Kortlandt, J. (1996) Kenya's flower exports, a flourishing business. In Kortlandt, J. and Sprang, U. (eds), *Make Way for Africa*. Amsterdam: InZet, Association for North-South Campaigns.

Durham, M. (1996) Western taste for prawns causes Third World misery. *The Observer*, 12 May, p. 15.

Freidberg, S. (2003) Cleaning up down South: supermarkets, ethical trade, and African horticulture. *Social and Cultural Geography*, 4, pp. 27–43.

Friedmann, H. (1993) The Political Economy of Food. *New Left Review*, 197, pp. 29–57.

Gereffi, G. (1994a) Capitalism, development and global commodity chains. In Sklair, L. (ed.), *Capitalism and Development*. London: Routledge.

Gereffi, G. (1994b) In Gereffi, G. and Korzeniewicz, M. (eds), *The Organization of Buyer-driven Global Commodity Chains: How US Retailers Shape Overseas Production Networks*. Westport, CT: Greenwood Press.

Green, D. (2001) Growing pains. *The Guardian*, 5 November, pp. 24–5.

Hale, A. (2000) What hope for 'ethical' trade in the globalised garment industry? *Antipode*, 32, pp. 349–56.

Hale, A. and Shaw, L. M. (2001) Women workers and the promise of ethical trade in the globalised garment industry: a serious beginning? *Antipode*, 33(3), pp. 510–30.

Hartwick, E. (1998) Geographies of consumption: a commodity chain approach. *Environment and Planning D: Society and Space*, 16, pp. 423–37.

Hartwick, E. (2000) Towards a geographical politics of consumption. *Environment and Planning A*, 32, pp. 1177–92.

Hughes, A. (2000) Retailers, knowledges and changing commodity networks: the case of the cut flower trade. *Geoforum*, 31, pp. 175–90.

Hughes, A. (2001a) Global commodity networks, ethical trade and governmentality: organising business responsibility in the Kenyan cut flower industry. *Transactions of the Institute of British Geographers*, 26, pp. 390–406.

Hughes, A. (2001b) Multi-stakeholder approaches to ethical trade: towards a reorganisation of UK retailers' global supply chains? *Journal of Economic Geography*, 1, pp. 421–37.

Jaffee, S. (1994) *Exporting High Value Food Commodities*. Washington, DC: World Bank.

Johns, R. and Vural, L. (2000) Class, geography, and the consumerist turn: UNITE and the stop sweatshops campaign. *Environment and Planning A*, 32, pp. 1193–213.

Kenya Flower Council (1998) *Code of Practice*. Nairobi: Kenya Flower Council.

Kenya Flower Council (1999) *Flowers from Kenya: the Newsletter of the Kenya Flower Council*, p. 4.

Klein, N. (2000) *No Logo*. London: HarperCollins.

Mackenzie, C. (1998) Ethical auditing and ethical knowledge. *Journal of Business Ethics*, 17, pp. 1395–402.

Maharaj, N. and Dorren, G. (1995) *The Game of the Rose*. Utrecht: International Books.

Marsden, T. and Wrigley, N. (1995) Regulation, retailing and consumption. *Environment and Planning A*, 27, pp. 1899–912.

Marsden, T., Flynn, A. and Harrison, M. (2000) *Consuming Interests: The Social Provision of Foods*. London: UCL Press.

Marsden, T. and Wrigley, N. (1996) Retailing, the food system and the regulatory state. In Wrigley, N. and Lowe, M. (eds), *Retailing, Consumption and Capital: Towards the New Retail Geography*. Harlow: Longman.

Meier, V. (1999) Cut flower production in Colombia – a major development success story for women? *Environment and Planning A*, 31, pp. 273–89.

Miller, D. (1987) *Material Culture and Mass Consumption*. Oxford: Blackwell.

Miller, D. (1998) Conclusion: a theory of virtualism. In Carrier, J. G. and Miller, D. (eds), *Virtualism: a New Political Economy*. Oxford and New York: Berg.

Miller, D. (2000) Virtualism – the culture of political economy. In Cook, I., Crouch, D., Naylor, S. and Ryan, J. R. (eds), *Cultural Turns/Geographical Turns: Perspectives on Cultural Geography*. Harlow: Prentice Hall.

O'Neill, O. (2002) *BBC Reith Lectures, 2002*, http://www.bbc.co.uk/radio4/reith2002/ (accessed 28 May 2002).

O'Riordan, T. (2000) Commentary: on corporate social reporting. *Environment and Planning A*, 32, pp. 1–4.

Orton, L. and Madden, P. (1996) *The Global Supermarket: Britain's Biggest Shops and Food from the Third World*. London: Christian Aid.

Pentland, B. (2000) Will auditors take over the world? Program, technique and the verification of everything. *Accounting, Organizations and Society*, 25, pp. 307–12.

Power, M. (1997) *The Audit Society: Rituals of Verification*. Oxford and New York: Oxford University Press.

Raynolds, L., Myhre, D., Figueroa, V., Buttel, F. and McMichael, P. (1993) The new internationalization of agriculture. *World Development*, 21, pp. 1101–21.

Renard, M. C. (1999) The interstices of globalization: the example of fair coffee. *Sociologia Ruralis*, 39, pp. 484–500.

Roddick, A. (2000) *Business as Unusual*. London: HarperCollins.

Shakespeare, J. (1995) Withering of the flower children. *The Observer*, 9 July, p. 16.

Stewart, S. (1994) *Colombian Flowers: the Gift of Love and Poison*. London: Christian Aid.

Strathern, M. (1997) 'Improving ratings': audit in the British university system. *European Review*, 5, pp. 305–21.

Watts, M. (1996) Development III: the global agrofood system and late twentieth-century development (or Kautsky *Redux*). *Progress in Human Geography*, 20, pp. 230–45.

Watts, M. and Goodman, D. (1997) Agrarian questions: global appetite, local metabolism: nature, culture, and industry in *fin-de-siecle* agro-food systems. In Goodman, D. and Watts, M. J. (eds), *Globalising Food: Agrarian Questions and Global Restructuring*. London and New York: Routledge.

Whatmore, S. and Thorne, L. (1997) Nourishing networks: alternative geographies of food. In Goodman, D. and Watts, M. J. (eds), *Globalising Food: Agrarian Questions and Global Restructuring*. London and New York: Routledge, pp. 287–304.

Wolf, J. (1997) Report on flower industry unearths danger to workers. *The Guardian*, 10 May, p. 16.

Zadek, S. (1998) Balancing performance, ethics, and accountability. *Journal of Business Ethics*, 17, pp. 1421–41.

12 The 'organic commodity' and other anomalies in the politics of consumption

Julie Guthman

Part of that quality stamp is also the guarantee of things which have not gone into the production process – a certification against adulteration, with the consumer able to enjoy the product safe in the knowledge that, thanks to the combination of environmental conditions and carefully regulated 'craft' production processes, they know precisely what they are eating or drinking. This guarantee has itself been generalised over whole areas of food production, most notably fruit and vegetable growing (but also animal husbandry), into the notion of organic foods. As an antidote to what many see as the sterility of the supermarket, there has been a blossoming of organic suppliers... Coming along as part of a growing consciousness about both environmental and health concerns over food production, organic food reinstates some of the attributes associated with wine regions: the labelling (sic) of food as organic tells consumers all about the conditions of its production (small-scale, chemical-free, non-intensive, locally sensitive, countercultural, etc.)

(Bell and Valentine, 1997, 155)

The idea that commodity chains have a geography places in the mind's eye widely dispersed nodes of production and intricate circuits of connectivity. Commodities are not made and consumed in one place, but composed and delivered in ways that are apparently elusive. What better justification for the proliferation of commodity studies research than to de-mystify one of the central ways in which globalization proceeds? Although writers draw upon a wide range of theoretical perspectives, there is a shared politics to this growing body of research: the politics of re-localization. That is, if the obscurity and intricacy of commodity circuits both enables and masks the systems of inequality upon which circuits depend, part of the solution is to re-localize them. It follows that re-localization can mean two different things. The first meaning is to bring commodity circuits into regional, downwardly scaled spaces. With an implicit emphasis on local self-sufficiency, this concept has drawn interest from Nazis to New Left utopianists in its complicated ideational history (see Bramwell, 1989 on Nazi and other fascist connections; cf. Kloppenberg *et al.*, 1996 on 'foodsheds'; Sale, 1985 as an example of a recent tract on bioregionalism).[1] A second, possibly more liberating sense is to give multiple commodity circuits 'place': to make them identifiable, connected and transparent. One might call it a search for origins.

However unlike the 'touristic quality' of a shallow multiculturalism that valorizes foreign or ethnic goods for their otherness, this is a search for origins that emphasizes 'biographies of production and distribution' (Cook and Crang, 1996, 146), the express purpose of which is to politicize global commodity chains.

The politics of re-localization is unusually pervasive in agro-food studies, despite arguments that the notion of globalization has less purchase on understanding the food sector than it does upon, for example, global automobiles (Goodman and Watts, 1994). Friedmann (1993) continues to indulge the importance of seasonality and locality to combat a food system based on distance and durability; Whatmore and Thorne (1997) wax rhapsodic about alternative networks of nourishment; Heffernan and Constance (1994, 48) see the local organization of food production as the second pole of a 'bipolar food system' where the first is a 'mass system of mega supermarkets hooked to TNC's'; and Bell and Valentine (1997, 194) view the 'foodie withdrawal from the global search for the exotic' as effectively a stance against globalization. In classrooms, following the food commodity is a well-regarded and widely used pedagogic tool among agro-food scholars. On the ground, 'eating local' and/or 'knowing where your food comes from' is a major discourse of alternative food provision. Besides the promises of community food security (read as self-sufficiency) in an era where consumers have heightened food anxieties, 'the anchoring of reputation in links of proximity' (Thevenot, 1998) can go to assuage such worries. In short, eating green (e.g. organic), eating ethically (e.g. Fair Trade) and eating locally – practices that Bell and Valentine (1997) group together – are distinct but analytically similar responses to increasing awareness of the ecological and social repercussions of international regimes of food production and distribution because they 'thicken' connections between producers and consumers (Crang, 1996).[2]

Implicit in this new politics of consumption is the Marxian notion of commodity fetishism, or the necessary masking of the social relations under which commodities are produced from which capitalist commodity production gains much of its legitimacy (see Hartwick, 1998). That society–*nature* relations are equally concealed in commodity production gives additional analytical purchase to this notion, especially in the case of land-based biological production such as agriculture (Allen and Kovach, 2000). Given food's status as 'the intimate commodity' (Winson, 1993) – the only commodity (along with medicine) that purposely and regularly passes through human bodies – the fact that many First World consumers do not know or do not care to know about the ecological origins of their food can only be construed as a commodity fetishism *par excellence*!

Understood in this way the objective of localizing or thickening becomes a means of opening up to scrutiny the social and ecological relations under which agricultural use values are created, as a first step towards transforming them. As Bell and Valentine (1997) suggest, food labelling becomes a way to provide heretofore concealed information about the materials and/or processes that agricultural producers incorporate and/or avoid. Hartwick (2000) also sees labelling as one of several approaches to de-mystifying the commodity.

By enabling a politics of consumption, labels in some sense defetishize commodities, yet this chapter questions the ability of labels to do this work in a substantive manner. While labels are the necessary ingredient to set so-called ethical commodities apart, by doing so labels allow civil protest and public choice to be conflated with consumption choice and profit-making. Moreover, by giving centrality to the commodity as vehicle of social change, they resurrect the fetishism of commodities in a back door way. Indeed, by indulging in their own mystifications they mask the geography of alternative commodity chains.

In the chapter, I primarily draw upon examples from the regulation of organic food.[3] 'Organically grown' is the most highly evolved of all eco-labels, and as Bell and Valentine (1997) suggest, organic production is often posed as *the* antidote to agro-food industrialization (and in popular parlance, globalization). Yet there are aspects of organic food provision itself that are industrializing and globalizing at a fairly rapid pace. This is most clearly the case in California, which, not coincidentally, has often been at the forefront of agricultural modernization. While there is a vibrant subsector of organic farmers whose practices and motivations are more in keeping with the organic imaginary, much organic production has been drawn into an oligosonistic industry structure that processes and distributes food to far-flung and impersonal markets. I argue that as the manifestation of a particular framework of regulation, the organic label has contributed to this outcome. Several further examples draw out the contradictions of labelling as a means to re-localize commodity chains and engender a politics of consumption.

The politics of consumption and the consumption of politics

Particularly because consumption – really consumption *choice* – has become central to today's political economy, it provides a specific axis for social change (Marsden and Arce, 1995). While writers such as Urry (1990) understand the new voluntarism and politicization of consumption choice to be a reaction to Fordist massification (cited in Warde, 1997); others such as Harvey (2000, 112), argue that it is the discretionary element of workers' disposable income that enables struggle over lifestyle and bodily practices equal to those over production. In either case, consumers increasingly have the opportunity to choose commodities that ostensibly are produced in ways that express moral preferences (Lipschutz and Fogel, 2003). If not necessarily conscious social activism, such reflexive consumption (DuPuis, 2000) has generated veritable shifts in demand.

Nowhere is this sort of reflexivity more evident than around food commodities. Against a background of escalating food politics, from the whimsical 'slow food movement' to highly visible and occasionally radical protests against genetically engineered organisms (GEOs), a sizable subgroup of consumers are now demanding agricultural products that do not involve those inputs and processes that historically have made agriculture profitable. Yet such a revitalized politics of consumption is not without contradictions. Although it does involve consumer intrusion into the productive sphere (Goodman and Watts, 1994) it is analytically distinct from consumer efforts to support the politics of production, as latter day boycotts have done. Rather, it tends to privilege the outcomes or externalities of production such as environmental quality and food safety (Lowe *et al.*, 1994) or to satisfy consumers' moral sensibilities. Even in the case of the anti-sweatshop movement, monitoring programs have been devised to satisfy consumers concerns and not to empower workers (Esbenshade, 2001).[4]

Furthermore, because of the centrality of consumption, the construction of food safety and quality is bound up with producers' constant search for value – or more accurately efforts by First World and multinational actors to *retain* value in a rapidly 'globalizing' economy. Indeed, if consumption involves consuming ideas, images and symbolic meanings (Bocock, 1993) there is 'a complex and shifting relationship between the two aspects of the use value of a commodity – its physical content and its interpretation' (Fine and Leopold, 1994, 26). Historically this gap between the commodity's (physical) use value and its imputed use value has been 'filled' by an 'aesthetic illusion' (Fine and Leopold, 1994): brand name. The material purpose of the aesthetic illusion is to widen the gap between realizable prices and actual costs of production (including a 'normal' rate of profit), that is, to create rent. Not only are rents eroded in periods of intense price competition, but also rent generation poses particularly interesting challenges when food is desired for freshness, rawness, wholeness, non-animal origin, nutritional composition, or the social and ecological conditions under which it was grown. Capitalist producers must devise ways to reincorporate these challenges to the food system (Beardsworth and Keil, 1997) and to solve the paradox of making profits by doing less the problem of value-added (Goodman and Redclift, 1991). This dilemma lies behind agro-industry's attempts to 'naturalize' global food provision by promoting fresh fruit and vegetables (Arce and Marsden, 1994).

The key strategy for stabilizing this gap and generating rents has become the use of signs, including voluntary labels. Yet an effective politics of consumption also requires signs to set valorized products apart (or signs to de-valorize in the case of boycotts), making the effect of voluntary labels highly ambiguous. While consumer knowledge about commodities allows consumers to exercise their preferences (Arce and Marsden, 1994, 303), such 'knowledges' are 'liable to be utilised in consumption as coinages in processes of cultural and social distinction' (Crang, 1996, 57). The 'geographic indications' attached to various agricultural products, such as appellations which explicitly codify meanings of tradition and authenticity and even valorize consumer tastes (Cook and Crang, 1996), brings consumers into circuits of regional development, at the same time that they are a form of localized value-seeking (Marsden, 1992). And while eco-labels provide rent-incentives for producers to incorporate more ecological practices, higher costs and class-constructed meanings around these products limit their proliferation. As Friedmann (1992, 86) explains:

> while privileged consumers eat free-range chickens prepared through handicraft methods in food shops, restaurants or by domestic servants, mass consumers eat reconstituted chicken foods from supermarket freezers or fast food restaurants and dispossessed peasants eat none at all.

In this last way, the ability to purchase relative freedom from risk (Beck, 1992) provides an exemplary case of neo-liberal regulation. The enhanced but highly differentiated consumer access provided by new 'rights to consume' (Marsden and Wrigley, 1995) not only plays an ideological role of conflating consumerism with freedom and citizenship, but also stimulates markets. Indeed what makes these labels unique from advertising claims is that they completely blur the line between social movement and business, as the regulation of organic labels has made clear.

De-mystifying organic products

Despite Bell and Valentine's (1997) portrayal of organic farming as small-scale, chemical-free, non-intensive, locally sensitive and countercultural, it is actually none of these. While there exist producers and/or production styles that fit one or more of these characterizations, the organic label does not guarantee any of them. The basis of this anomaly is a profound tension between the organic movement and the organic industry, between a particular vision of public good and private profit-making.[5] This tension was made manifest in why, what, and how organic meanings came to be regulated. In short, the organic industry pursued a label to set organic food apart for purposes of trade. Along the way, those involved made a number of politicized decisions that constructed organic meanings in a narrow and technical sense. The institutions and conventions developed to support these meanings encouraged rent-seeking, which had profound impacts upon the constitution of organic commodity chains.

In the interest of trade

The organic movement itself had heterogeneous philosophical, historical, and constituent roots which were brought together in an uneasy alliance by the counterculture of the late 1960s and early 1970s. At that time no one meaning was attributed to organic agriculture. If anything, the organic movement was characterized by its holism, borrowing promiscuously from several different historical strains and thriving on amorphous meanings. Nevertheless the marked influence of the New Left implicitly linked organic agriculture with critiques of industrialized food production. Organic agriculture and other alternative systems of food delivery that were developed simultaneously had an explicitly radical tone in the early 1970s, and many early practitioners and consumers embraced organic food precisely because it existed outside of the conventional food system.

Even so, this stance from the margins was always ambiguous. First, even radical social movements need to grow to demonstrate their force. Part of the US movement was led by the pragmatic Rodale family.[6] Hardly sceptical of capitalist institutions, the Rodales sought to spread the techniques of organic farming in part to expand their sphere in the magazine publishing world. Business-oriented growers similarly felt they could benefit financially from the expansion of organic consumption. Finally, there was a real need for consumers to obtain assurances about the provenance of their purchases. Minimally, the claim of 'organically grown' meant many different things to many different people and there were occasional cases of fraud. The Rodales were the first to take an interest in institutional procedures to deal with fraud, which they blamed on short supplies and lax regulation (Nowacek, 1997). In 1972, they initiated their own certification programme for organically grown food, the Rodale Seal of Approval. Most of the organic farmers they dealt with were in California, and the Rodale certification program shortly evolved into the California Certified Organic Farmers (CCOF), founded in 1973.

In the beginning, CCOF was an unstructured group of about fifty mostly hippie farmers, who showed notable resistance to any sort of institutionalization. It was not until the late 1980s, for example, that the CCOF received tax exempt status from the federal government and filed their first tax returns. As farmers, the common interest among the first CCOF members was in developing, refining, and sharing a set of *production* practices. A particular group of CCOF growers sought to recognize these practices in the market, both to protect consumers from false claims and to differentiate the quality of their product in an overt way. This subgroup pushed for uniform definitions and standards and a certification programme to verify the practices of member farms – although it should be noted that certification was appreciably more informal than it is today. Their efforts were joined by like organizations in other states, starting with Oregon Tilth; by the end of 1974, eleven other regional certification organizations had been formed, largely with the support of Rodale's *Organic Farming and Gardening* magazine (Nowacek, 1997).

Having established a modicum of legitimacy, CCOF went on to pursue legislative avenues to define and enforce the meaning of organic, an objective that was contested among those in its membership who distrusted government involvement. Nevertheless, with CCOF's persistence, California passed an Organic Food Act in 1979. Over the next decade demand for organic products and use of the organic label expanded rapidly. North American consumer 'scares' concerning residues left by toxic materials such as Alar and Aldicarb gave many growers the incentive to become instantly organic and fraudulent uses of the term became increasingly visible. CCOF's Board again voted to take decisive action which culminated in the California Organic Foods Act of 1990 (COFA). The COFA established a legal baseline definition of organic growing practices, including a list of materials allowed to be used in organic production systems. It became the model for a national standard and for the first time included enforcement provisions. Meanwhile, CCOF continued to be a standard-setter as well as becoming highly regarded in international circles. Today it certifies far more California growers than any other certifier.

Paralleling these state efforts, there was a good deal of activity at the federal level to establish organic meanings. Actors at the national level were particularly fearful that the word 'organic' would become meaningless as had occurred when 'natural' was co-opted by the major food companies. In 1984 the first nationwide organic trade group (later to become the Organic Trade Association) was formed. Composed primarily of early certification associations and larger producers, but working with consumer advocacy groups, its stated goal was to develop a national organic labelling law to resolve conflicts in state laws and hence ease trade (Mergentime, 1994). In other words, those most desirous of a federal law were the ones who were engaged in interstate trade, where uniformity of standards is critical. Similarly, federal regulation was supported by companies who dealt with processed food items, since the most contentious legal issues were tangential to fresh fruit and vegetable production. Implicitly, producers expected to define organic in ways that would continue to give it distinction, but at the same time move it into the mainstream. By 1990, sufficient support was obtained to pass the federal Organic Foods Production Act (OFPA), although the Act's implementation was held up for more than ten years in an ongoing struggle between the US Department of Agriculture (USDA) and various organic movement constituencies.

Thus as the organic movement increasingly became led by third party certifying agencies and producers' associations, 'organically grown' came to be defined specifically as a production standard for farmers (and later processors) rather than as a food safety standard for consumers. Certainly it did not represent an alternative system of food provision. The organic movement thereafter evolved into a drive for institutional legitimacy and regulation of the term 'organically grown' in the interests of trade. While not intending to create a label *per se*, the movement effectively ensured that the right to claim that any product is organically produced was contingent upon compliance with legal definitions, which included a set of labelling requirements. So although codification arose from multiple intentions, its greatest success was to open up markets.

For simplicity and transparency

The definition of organic agriculture currently promoted by CCOF and recently adopted by the National Organic Standards Board (which advises the OFPA) is as follows:

> an ecological production management system that promotes and enhances biodiversity, biological cycles and soil biological activity. It is based on minimal use of off-farm inputs and on management practices that restore, maintain and enhance ecological harmony. The principal guidelines for organic production are to use materials and practices that enhance the ecological balance of natural systems and that integrate the parts of the farming system into an ecological whole.
>
> (CCOF, 1998, 6)

This definition is the product of many years of debate about what organic should mean. It presents an agroecological ideal that does not easily fit into practicable standards, nor does it make any guarantees about outcomes. It does provide guiding principles for organic production, however given the way in which meaning is enforced, organic's main thrust is the avoidance of certain classes of inputs. Its social agenda, moreover, is truly vague.

This narrow and technical definition is a product of rule-making itself. Both in practical and political terms, the term organic was deliberately circumscribed. In practical terms, the task of rule-making was to establish enforceable standards. Those involved faced the enormous difficulty of melding contested and sometimes contradictory imperatives into a single standard and bounding the issues to be addressed. For example, there exists a clear tension between incorporating local specificity – seen as crucial to agroecology – and creating a broadly recognized and uniform standard. In the interest of consumer transparency and producer reciprocity, regulators continue to strive for uniformity. In political terms, there were contradictory imperatives to define standards in ways that protected existing participants but at the same time were not unduly exclusive. 'Organically grown' needed to be reduced to a technical term (cf. Allen and Sachs, 1993; Buttel, 1993) so that anyone could participate, if not necessarily on their own terms.

Thus while it may be the case that organic standard-setters did not intend to incorporate a substantive critique of conventional agro-food delivery, politicized decisions made along the way narrowed the social focus of organic. One of the formative issues was that of scale. Despite the theoretical (and practical) difficulties of codifying a particular scale of operation, organic production is idiomatically linked with small-scale, populist agrarianism. From time to time, activists introduced the idea of imposing an acreage limitation on those within the CCOF program, but the motion was never successful.[7] Neither were proposed standards addressing labour conditions and remuneration, even though labour issues, too, remain a discursive allure. Other ideas circulating within organic discourse also have received little institutional bite. Some organic certifiers have as a guiding principle support for regional food systems, but there are no standards to support such efforts. The idea of minimal processing was once a mainstay of the concept of natural foods but has been abandoned as all the major certifiers have become involved in certifying food processors. Likewise rhetorical nods are often made to the fair economic return obtained by producers, but certification programmes lack any enforcement mechanisms. In short, 'organically grown' has been reduced to a standard of production practices alone.[8]

In addition to this definitional narrowing to a focus on production, organic meanings have been further constrained by another set of forces involving actors, as well as actants. The original idea behind a production standard for organic was to codify processes that replicate those found in nature, and the lack of marketable organic inputs reinforced an emphasis on shared practices. Unfortunately, processes and practices have been difficult to regulate. Despite the mild exhortations by certifiers that farmers should adopt plans that incorporate biological controls and soil development, the focus of regulation gradually shifted to inputs and materials, concomitant with the growth of an organic input market and organic production itself. Further, the first legal definition of organically grown allowed only the use of materials found 'in nature' and prohibited the use of synthetically produced fertilizers, pesticides, herbicides and other inputs. Yet such a seemingly straightforward definition created its own set of problems. For example, it raised the question as to whether botanical pesticides found in nature but which create 'pesticide treadmills' similar to those established by chemically derived pesticides, should be considered organic.[9] In practice materials were assessed on a case-by-case basis – often with significant contestation – but always with reference to a more fundamental distinction between 'natural' and 'synthetic' products.[10] The crux of organic regulation thus became the so-called materials list, which itemizes and differentiates between allowable, restricted, and prohibited inputs for organic farming.

To create incentives

Like all appellations, the organic label depends upon the construction and maintenance of quasi-monopoly conditions to give it market meaning. In part by design and in part by default, comprehensive state involvement was largely absent from processes of external verification that organic foods were produced in accordance with established standards until recently. Private third-party certification became the primary mode of establishing these conditions.[11] Certification became a virtual necessity for interstate commerce in organics and from 2002 was required by a new federal law.[12]

To be certified, growers must complete elaborate paper work including a farm plan; agree to initial, annual and perhaps spot inspections; fulfil whatever requirements for crop or soil sampling, pay various dues, fees and assessments; and, of course, agree to abide by the practices and input restrictions designated by that agency and the law itself. Enforcement usually involves action taken by the certifier. Forms of censure include prohibiting a certain crop to be sold as certified organic (in the case of drift violations), fines and de-certification. In all cases the burden of proof rests almost entirely on the alleged violator, and there are varying degrees of due process in making that burden.

While some state laws require that certification be pursued through state departments of agriculture, most certification occurs through private agencies.[13] Most of the older certification agencies are non-profit organizations that developed through the organic movement and thus also tend to be advocacy-oriented. While varying degrees of democratic decision-making exist, these organizations tend towards self-regulation in that growers make decisions and set standards and there often is peer review of operations.[14] As such, standards tend to be set according to what represented growers want and need and are not necessarily defined in some wider public sense. At the same time, non-profit certifiers also function as trade organizations and thus have much at stake in upholding the integrity of organic food. Enforcement actions are relatively strong and the achievement of requisite organic standards is comparatively difficult. Since the late 1980s, another type of private certifier has emerged. These are private for-profit organizations that treat certification as a business service, removing them from the politics of defining organic. Instead, they trade on the ease of their certification processes; the for-profit organizations tend to require less paperwork, limit their inspections, and certify within a week of two of application. They also tend to be less transparent in their practices: some (but not all) do not make their standards or certified growers publicly available.

Certification exists both to protect consumers from fraudulent claims and to reward producers who conform to particular practices. The most unmistakable barriers to entry are associated with certification itself. Fees and compliance costs can be imposing; bureaucratic hassle and being subject to unusual levels of surveillance also act as disincentives to entry. The key barrier, however, is the required transition period when crops cannot be harvested and sold as organic unless three years have passed since the last disallowed substance was applied to the land on which they are grown. During these three years, yields generally decline as prohibited substances are abruptly withdrawn but biological controls have not had adequate time to be established. In the intervening period crops must be sold at conventional prices, thereby imposing substantial opportunity costs. While the explicit purpose of the transition is to minimize toxic residues in the soil, the implicit purpose is to prevent rapid entry into the organic market. Although some growers are able to skirt these barriers by (for example) bringing marginal fallow land into production, many also find it too challenging to grow in compliance with organic definitions. In several ways, then, certification acts to create scarcity – the basis of the much sought-after price premium.

Constituting organic commodity chains[15]

All of the conventions of organic regulation have important implications for the resultant shape of the organic commodity chain. Creating a defined organic sector is not just a matter of codifying a system of recognition and reward for an already existing group of producers, as might be the case with other appellations. Rather, its putative design has been to provide a financial incentive to any grower who adopts organic techniques. So, for instance, in Guthman (2000), the most oft-cited reason for which growers sought to convert to organic production was the achievement of specific price premiums or the use of cropping strategies that net more value per acre. Many more growers were brought in by buyers, who themselves have strongholds in the organic market. At the same time, new competition is eroding the very premiums organic products might achieve, and is beginning to reward economies of scale.[16]

Nor does the organic label preclude certain types of producers from participating. The regulatory focus on inputs and materials, along with the growth of an organic input market has reinforced the idea that soil fertility and pest control are things that can be bought. So, in spite of the rhetoric that organic farming is based on minimal use of off-farm inputs, growers increasingly adopted strategies of 'input substitution' (Rosset and Altieri, 1997). Particularly in the case of growers converting to organic production, allowable substitutes were used to replace materials normally used in conventional production.

As a result, organic production does not live up to many of its myths. It is true that most Californian organic farms are much smaller than conventional farms (Klonsky and Tourte, 1998; cf. Department of Finance, 1997), yet many holdings are hobby farms or tax write-offs for residential real estate. Meanwhile, some of the largest agribusiness farms in California are involved in organic production, instantiating organizational forms that have been the cause of long-standing critiques of industrial agriculture (see McWilliams, 1971 [1935]). More importantly, the industrial structure of the sector is veering toward oligopsony in the sense that a few very large firms are marketing for – and thus increasingly controlling – many others.

In terms of spatial scaling, organic commodity chains are not necessarily local nor foreshortened. While the organic sector includes a vibrant direct marketing subsector, many growers are involved in production for export. Further, knowledges across the chain can be limited – even amongst those involved in production. In Guthman (2000), 17 per cent of growers interviewed know their crops reached export markets; whilst 21 per cent in a national survey were aware of such geographies (OFRF, 1999). A lack of transparency across organic chains also raises questions about the possibility for consumers to 'thicken' their relationships with producers (cf. Crang, 1996). In the Guthman study, 51 per cent of producers said they sell at least some of their crops to shippers or processors on contract, and 44 per cent said they sell in wholesale markets.

Other labelling efforts

Organic labelling is just one example of many ongoing attempts to re-localize food production. Others, such as wine appellations and labels specifying craft-production, are juxtaposed to mass production and valorized for taste and their contribution to regional economies but make no bones about catering to elite sensibilities. Nor do such labels normally suggest promises of social change. Labels that are more explicitly political can be more problematic.

At one end of the spectrum are eco-labels such as The Food Alliance-approved. While promising to go 'beyond organic' in its concerns for workers and small scale family farmers, and allowing a gradualist approach, The Food Alliance standards are in fact quite nebulous. Farms are scored using an evaluation tool 'specifically designed with the goals of The Food Alliance in mind' (The Food Alliance, 2000). Farmers must obtain a 70 per cent threshold on this evaluation tool to be eligible and must also submit regular farm improvement plans. Moreover, vague prescriptions like the use of 'least toxic methods' form part of the evaluation. Unlike organic labelling, the Food Alliance started from the identification of a market segment by a marketing group (Hartman Group, 1997) and then developed production standards to capture that market (Coody, 1999). Here the explicit purpose is to generate value. Other eco-labels such as Cafe Mam, an organic coffee producer whose label also promises social responsibility, similarly fail to make transparent the conditions under which the product was made (Goodman, 1999).

At the other end of the spectrum are various labels which declare that products are free from genetically modified (or engineered) organisms (GMOs or GEOs). These labels are largely unregulated by either state or national governments, although the Food and Drug Administration does limit the language used so as not to disparage other producers. Should such labels be allowed to proliferate, producers will no doubt try to cash in on any economic rents to be had, although ironically their purpose is *not* to generate a price premium. Rather, the central intent of declarations that products are GMO- or GEO-free is to pressure producers and regulators to reconsider the need and efficacy of genetic modification in agricultural production. While I have discussed the function of these labels as a form of regulatory privatization elsewhere (Guthman, 2003) their impact has already proven to be broad, as evidenced by the several multinational food firms that have recently pledged to avoid the use of genetically engineered ingredients. Moreover, to the extent to which labelling has made the potential risks of agricultural biotechnologies a subject of everyday discourse, organizations which support such labelling *have* opened up to scrutiny the social and ecological relations under which agricultural use values are produced.

The limits of labelling

Bell and Valentine (1997) have suggested that the search for the exotic in food choices is simultaneously a search for the authentic, as both are based on sophisticated knowledge. While it is possible that such knowledge serves to thicken connections between producers and consumers, attempts to secure additional knowledge through labelling may simply re-fetishize alternative meanings. In the case of organic food labelling, the industry plays upon the organic movement's connotations in ways that border on the cynical, as organic agriculture does not necessarily provide a systemic alternative to conventional commodity chains. Indeed, the organic label creates its own fictions and erasures in order to fill the gap between physical use value and signification. So, for instance, a Health Valley organic cereal box tells consumers that 'by supporting dozens of small organic farms, Health Valley helps protect a way of life for these family farmers, and helps ensure a safer and healthier Earth for you and your children.' Since Health Valley is a major producer and distributor of both organic and non-organic processed food items, at best this is what Hartwick (2000) might call a 'partial admission', a biased and incomplete description of the actual commodity chain.

Given that organic's entry barriers are somewhat permeable, the real success of organic regulation is the discursive work it does. Not only does the organic designation inform consumers that organic food may be safer and/or environmentally protective – characterizations which arguably have an empirical basis, it also inscribes significance in areas where organic falls short: from 'saving' the family farm, to providing a better working environment, to provisioning whole and/or more nutritious food. Organic labelling has the additional advantage of having a 'movement' behind it, ultimately bestowing more meaning and durability than a brand name might. In that way, it, fetishizes the process of social change itself, by suggesting that purchasing a commodity is sufficient to effect such change. If organic food was truly an antidote to processes of commodification, the 'organic commodity' surely would be seen as an oxymoron.

Of more immediacy is the fact that the construction of alternative meanings encourages rent-seeking, undermining the very goals such labels seek to promote.[17] In the organic case, rents attract entry into the sector, creating the competition that eventually erodes such rents. As a result there is a perverse tension between upholding, or even strengthening barriers to entry and making it easier for growers to convert. The former position is voiced by those in the movement who seek to hold new entrants to the ideals that organic is supposed to instantiate; the latter position is voiced by those in the industry who want to proliferate ecologically sounder practices – albeit more narrowly defined – in the interest of expanding market share. Nevertheless, those who seek to revitalize, or even deepen organic meanings, are also looking for self-protection in an increasingly competitive market. And they crucially depend on consumers to pay the difference. As an antidote to growing doubts expressed about the socio-economic and environmental conditions of mainstream agro-food systems, preciousness is a dubious solution.

Acknowledgements

The author wishes to thank Alex Hughes and Suzanne Reimer for useful editorial comments. Research was in part supported by grants from the National Science Foundation (SBR-9711262), the University of California's Sustainable Agriculture Research and Education Program, and the Association of American Geographers.

Notes

1 Furthermore, as McKenna and Campbell (2000) assert, this politics of scale assumes a clear and fixed ontological mapping between spatial scales and particular political and ideological positions.

2 Just as globalization is a gloss for discrete and occasionally counter-veiling processes within advanced capitalism, so is localization a gloss for a wide range of purportedly contrarian food practices.

3 The portions of this chapter that specifically address organic production and regulation are drawn from the author's dissertation research (Guthman, 2000). The study included over 150 semi-structured interviews with both all-organic and mixed (i.e. both conventional and organic) growers that took place in 1998 and 1999 in California. Data on organic regulation were collected through in-depth interviews with representatives from several of the certifying agencies who operate within California, along with public officials, advocates, and technical experts. In addition, the study included textual analysis of legislative and certifier archives, industry conferences, and on-line discussions.

4 Hartwick (2000, 1182) sees this limited sort of reflexivity mirrored in academia where geographic theories of consumption 'project a guilty conscience about over-consumption' but say 'next to nothing on labour'.

5 The distinction between movement and industry is more an analytical divide than a clear mapping of different actors. Not only are many inclined to mix movement and industry perspectives in any given discourse, but also distinctions between the two are already muddied by the fact that most organic producers sell their products and are subject to at least some market logic. For these complications, those involved often refer to themselves as the 'organic community'.

6 The Rodale family started the first experimental farm in the US devoted explicitly to organic production, located in Emmaus, Pennsylvania. They also published several books and magazines on organic farming and holistic health, the most well known of which were *Organic Farming* and *Gardening and Prevention*, respectively.

7 Some certifiers do impose acreage constraints on participating producers and others have as a guiding principle that organic operations should be small scale.

8 There are quite specific standards for handling and marketing, in regard to labelling and segregation of organic product, but these serve mainly to uphold the abstraction of 'organically grown'.

9 It also raised the more fundamental problem that the use of 'natural' or 'found in nature' as the basis of acceptability relied on what are clearly problematic assumptions about the essential goodness of nature (Cronon, 1995; Williams, 1980; cf. Jackson, 1980) and further made all decisions subject to the infinite regress of the nature/culture divide.

10 The California Organic Foods Act 1990, for example, prohibits 'synthetically compounded fertilizers, pesticides and growth regulators', where 'synthetically compounded' refers to 'a process which chemically changes a material extracted from naturally occurring plant, animal, or mineral sources, excepting microbiological processes'. Yet it also prohibits some toxic materials found in nature (e.g. strychnine) and allows for certain synthetic materials.

11 Some states also prohibit substitutes, restricting the use of terms such as 'ecologically grown'.

12 Until the federal law's implementation in 2002 certification of growers varied according to different state laws. California, for example, allowed growers simply to register with the state to sell products as organic. Verification and enforcement was accordingly weaker.

13 State certification often is criticized from within the industry for being unresponsive and bureaucratic. While these accusations are no doubt true, state enforcement is also more legally circumscribed. States have a constitutional obligation to ensure due process, whereas private entities have much more discretion in these matters. It is telling that organic producers (for the most part) prefer the politics of private certification, one of the reasons why they resisted the federal law.

14 Although they refer to themselves as third party certifiers, trade associations that do most organic certification can only offer second party certification, with third party status reserved for organizations completely independent of the industry they certify (Caldwell, 1988; cited in Lipschutz and Fogel, 2003). With full implementation of the federal law, all certifiers will have to prove third party distance.

15 The plural is used both because crop specificities matter tremendously in organic production, due to the uneven availability of efficacious inputs, and the fact that some organic commodity chains involve a very diverse crop mix.

16 This is where organic substantially differs from regional appellations, which have a ready-made spatial monopoly, if value can be constructed around the territory. Yet even the best wine regions face competition when new wine regions receive accolades and can compete with lower costs.

17 It should be noted that relatively higher prices for organic production do not entirely comprise rents. With both the investment costs and the difficulty of growing certain crops, organic can indeed cost more than conventional food

production, especially if growers de-intensify their cropping strategies to include marginal value crops – which few do in California. Moreover, when they do exist, these rents can potentially provide the cushion to address some of the social justice concerns that have thus far been substantially neglected. The problem is when rents remain 'excess profits', or are shifted to (even created within) other nodes in the commodity chain such as retail sites. This is a phenomena which many in the organic sector already have observed.

References

Allen, P. and Sachs, C. (1993) Sustainable agriculture in the United States: engagements, silences, and possibilities for transformation. In Allen, P. (ed.), *Food for the Future*. New York: John Wiley and Sons.

Allen, P. and Kovach, M. (2000) The capitalist composition of organic: the potential of markets in fulfilling the promise of organic agriculture. *Agriculture and Human Values*, 17, pp. 221–32.

Arce, A. and Marsden, T. (1994) The social construction of international food: a new research agenda. *Economic Geography*, 69, pp. 293–311.

Beardsworth, A. and Keil, T. (1997) *Sociology on the Menu*. London: Routledge.

Beck, U. (1992) *Risk Society: Towards a New Modernity*, Ritter, M., translator. Thousand Oaks: Sage Publications.

Bell, D. and Valentine, G. (1997) *Consuming Geographies: We Are Where We Eat*. London: Routledge.

Bocock, R. (1993) *Consumption*. London: Routledge.

Bramwell, A. (1989) *Ecology in the Twentieth Century: A History*. London: Yale University Press.

Buttel, F. H. (1993) The production of agricultural sustainability: observations from the sociology of science and technology. In Allen, P. (ed.), *Food for the Future*. New York: John Wiley and Sons.

Caldwell, D. J. (1998) Ecolabeling and the regulatory framework – a survey of domestic and international fora. Consumer's Choice Council, www.consumerscouncil. org/ccc/, October 30.

CCOF (1998) Certification Handbook, California Certified Organic Farmers, Santa Cruz.

Coody, L. (1999) Scrutinizing labels or eco-labelling: clean, organic, sustainable? Presented at *19th Annual Ecological Farming Conference*, Asilomar, CA.

Cook, I. and Crang, P. (1996) The world on a plate: culinary culture, displacement and geographical knowledges. *Journal of Material Culture*, 1, pp. 131–54.

Crang, P. (1996) Displacement, consumption and identity. *Environment and Planning A*, 28, pp. 47–67.

Cronon, W. (ed.) (1995) *Uncommon Ground: Toward Reinventing Nature*. New York: W. W. Norton and Company.

Department of Finance (1997) *California Statistical Abstract*, 38th Edition, State of California, Sacramento.

DuPuis, M. (2000) Not in my body: rBGH and the rise of organic milk. *Agriculture and Human Values*, 17, pp. 285–95.

Esbenshade, J. (2001) The social accountability contract: private monitoring and labour relations in the global apparel industry. PhD dissertation, Ethnic Studies, University of California, Berkeley.

Fine, B. and Leopold, E. (1994) *The World of Consumption.* London: Routledge.

Friedmann, H. (1992) Changes in the international division of labour: agri-food complexes and export agriculture. In Friedland, W. H., Busch, L., Buttel, F. H. and Rudy, A. P. (eds), *Towards a New Political Economy of Agriculture.* Boulder: Westview Press.

Friedmann, H. (1993) After Midas's feast: alternative food regimes for the future. In Allen, P. (ed.), *Food for the Future.* New York: John Wiley and Sons.

Goodman, M. (1999) Developmental consumption: embedding relationships of the international organic food commodity. Paper presented at Conventional and Organic Agriculture – Encounters at the Interface, University of California, Santa Cruz.

Goodman, D. and Redclift, M. (1991) *Refashioning Nature.* London: Routledge.

Goodman, D. and Watts, M. (1994) Reconfiguring the rural or fording the divide. *Journal of Peasant Studies,* 221, pp. 1–49.

Guthman, J. (2000) Agrarian dreams? The paradox of organic farming in California. PhD dissertation, Department of Geography, University of California, Berkeley.

Guthman, J. (2003) Eating risk: the politics of labelling transgenic foods. In Kelso, D. and Schurman, R. (eds), *Recreating the World: Genetic Engineering and its Discontents.* Berkeley: University of California Press.

Hartman Group (1997) The Hartman report: food and the environment, a consumer's perspective. The Food Alliance, Portland, OR.

Hartwick, E. (1998) Geographies of consumption: a commodity-chain approach. *Environment and Planning D: Society and Space,* 16, pp. 423–37.

Hartwick, E. (2000) Towards a geographical politics of consumption. *Environment and Planning A,* 32, pp. 1177–92.

Harvey, D. (2000) *Spaces of Hope.* Berkeley: University of California Press.

Heffernan, W. D. and Constance, D. H. (1994) Transnational corporations and the globalization of the food system. In Bonanno, A., Busch, L., Friedland, W., Gouveia, L. and Mingione, E. (eds), *From Columbus to ConAgra.* Kansas City: University Press of Kansas.

Jackson, W. (1990) Agriculture with nature as analogy. In Francis, C. A., Flora, C. B. and King, L. D. (eds), *Sustainable Agriculture in Temperate Zones.* New York: John Wiley and Sons.

Klonsky, K. and Tourte, L. (1998) Statistical review of California's organic agriculture 1992–1995. Davis: University of California Agricultural Issues Center.

Kloppenberg, J. J., Henrickson, J. and Stevenson, G. W. (1996) Coming into the foodshed. *Agriculture and Human Values,* 13, pp. 33–42.

Lipschutz, R. D. and Fogel, C. (2003) Regulation for the rest of us: global civil society and the privatization of transnational regulation. In Biersteker, T. J. and Hall, R. B. (eds), *The Emergence of Private Authority: Forms of Private Authority and Their Implications for Global Governance.* Cambridge: Cambridge University Press, pp. 115–40.

Lowe, P., Marsden, T. and Whatmore, S. (1994) Changing regulatory orders: the analysis of the economic governance of agriculture. In Lowe, P., Marsden, T. and Whatmore, S. (eds), *Regulating Agriculture.* London: David Fulton.

McKenna, M. and Campbell, H. (2000) It's not easy being green: 'Food Safety' practices in New Zealand's apple industry. Paper presented at X World Congress of Rural Sociology, Rio de Janeiro, Brazil.

McWilliams, C. (1971 [1935]) *Factories in the Field.* Santa Barbara: Peregrine Smith.

Marsden, T. (1992) Exploring a rural sociology for the fordist transition. *Sociologia Ruralis*, XXXII, pp. 209–30.

Marsden, T. and Arce, A. (1995) Constructing quality: emerging food networks in the rural transition. *Environment and Planning A*, 27, pp. 1261–79.

Marsden, T. and Wrigley, N. (1995) Regulation, retailing and consumption. *Environment and Planning A*, 27, pp. 1899–912.

Mergentime, K. (1994) History of organic. *Natural Food Merchandiser's Organic Times*, pp. 62–6.

Nowacek, D. (1997) The organic foods system from 1969–1996: a defence of associative order and democratic control over a market. MA thesis, Rural Sociology, University of Wisconsin, Madison.

OFRF (1999) Third Biennial National Organic Farmers' Survey final report of survey. Organic Farming Research Foundation, Santa Cruz.

Rosset, P. M. and Altieri, M. (1997) Agroecology versus input substitution: a fundamental contradiction of sustainable agriculture. *Society and Natural Resources*, 10, pp. 283–95.

Sale, K. (1985) *Dwellers in the Land: The Bioregional Vision*. San Francisco: Sierra Club Books.

The Food Alliance (2000) www.thefoodalliance.org

Thevenot, L. (1998) Innovating in 'qualified' markets: quality, norms, and conventions. Paper presented at Workshop on Systems and Trajectories of Agricultural Innovation, Institute of International Studies, UC Berkeley.

Urry, J. (1990) *The Tourist Gaze: Leisure and Travel in Contemporary Societies*. London: Sage.

Warde, A. (1997) *Consumption, Food and Taste*. London: Sage.

Whatmore, S. and Thorne, L. (1997) Nourishing networks: alternative geographies of food. In Goodman, D. and Watts, M. J. (eds), *Globalising Food: Agrarian Questions and Global Restructuring*. London: Routledge.

Williams, R. (1980) *Problems in Materialism and Culture*. London: Verso.

Winson, A. (1993) *The Intimate Commodity: Food and the Development of the Agro-Industrial Complex in Canada*. Toronto: Garamond Press.

13 Knowledge, ethics and power in the home furnishings commodity chain

Suzanne Reimer and Deborah Leslie

Introduction

Much popular discussion of commodities increasingly has been concerned with political and ethical questions surrounding the origins and production of goods (Klein, 2000). Individuals, consumer groups, nation states and supra-national bodies such as the European Union have sought to interrogate the nature of production–consumption relations (Burgess, 2001). Political debates have centred upon the problematics surrounding 'food safety' for consumers, the implications of the cultivation and consumption of genetically modified organisms, and the pernicious nature of labour processes central to low-cost production of clothing and footwear sold to Northern consumers at high prices. Notions of audit have become important as consumers seek increasing knowledge about the production of the food they eat, the clothes they wear and the objects they purchase for their home.

Flowing from – and bound up with – these political concerns, there has been growing academic interest in the politics of consumption. Attention has focussed upon the significance of consumer boycotts, corporate campaigns, product labelling and fair and ethical trade initiatives (Hale and Shaw, 2000, 2001; Hartwick, 1998, 2000; Hughes, 2001, Chapter 11, this volume; Johns and Vural, 2000). Although ethical consumption debates and analyses of boycotts have a longer history outside geography (e.g. Smith, 1990), recent discussions within the discipline can be seen to derive from two closely related sets of literatures. The first reflects a growing interest in the practices and knowledges of consumption (Cook and Crang, 1996; Crang, 1996; Jackson, 1999, 2002; Miller, 1995). Second, writers have begun to consider the mutual dependence of production, retailing, design, advertising and final consumption. While some analysts have emphasized systemic links between nodes in a commodity chain, others have conceptualized flows of goods and meanings through a more amorphous 'network' or 'circuit' (Hughes, 2000; Leslie and Reimer, 1999).

Prompted by debates surrounding consumer knowledges (Cook and Crang, 1996) as well as the growing literature on the importance of producer knowledges to reconfigurations of commodity chains and networks (Crang, 2000; Hughes, 2000; Morgan and Murdoch, 2000), this chapter foregrounds the politics of furniture consumption. We are particularly concerned with the extent to which home furnishings consumption choices may be driven by knowledges of conditions at other sites in the chain or network. Our intent is explicitly to interrogate consumer knowledges, although we also are attentive to the ways in which these understandings are shaped by the intentions and actions of retailers, manufacturers and designers of furniture.[1]

We reflect upon awareness of the following aspects of furniture production: materials used; environmental consequences of forestry practices and employment conditions within the industry. We thus group together a relatively broad range of concerns relating to what may be loosely referred to as 'ethical consumption'. There is some danger of conflating the dynamics of different political movements and concerns. As Hughes (2001, 390) notes, 'ethical' initiatives developed to ensure minimum environmental and labour standards contrast markedly with 'fair trade' projects which seek more explicitly to increase producers' control over supply chains. The agendas of environmentally friendly consumption might not always coincide with a progressive labour politics – as in the tendency for Californian organic agriculture to rely upon 'nonunionised, casual migrant wage labour' (Goodman, 2000, 216). It also is important to highlight the geographies of ethical consumerism. At a global level, Raynolds (2000, 67), suggests that European fair trade movements historically have been most concerned to provide alternatives to 'imperialist tendencies in international commerce', while consumer consciousness in the United States predominantly has been concerned with environmental and food safety rather than social justice issues.

Existing evaluations of the potential for consumption practices to reconfigure power relations along the commodity chain predominantly have focussed upon the food and clothing industries (Hartwick, 1998, 2000; Johns and Vural, 2000; Murdoch *et al.*, 2000). However the distinctive spatialities and temporalities of furniture make it a rather different type of commodity. We begin with an overview of the politics of consumption in the home furnishings sector, drawing attention to moments when ethical issues may come to the fore. We then suggest that furniture's relatively distant relationship with the body as well as its comparative longevity have important implications for consumers' ethical decision-making processes. Finally, we reflect upon the role of the prominent 'lifestyle' retailer Ikea in configuring possibilities for the environmentally sympathetic consumption of home furnishings.

In an earlier review of theoretical approaches to commodity chain dynamics (Leslie and Reimer, 1999), we noted that while our general sympathies tended towards circuit- or network-based analyses, retaining some notion of where power lies in a commodity chain is required in order to ground political action (see also Hale and Shaw, 2001). Our reading of the home furnishings chain as characterized by multi-stranded webs of interaction between people, resources (trees, minerals, plants), chemicals, technologies and habitats has been informed by examinations of the agro-food sector which have adopted actor-network theoretical approaches (Goodman, 1999; Whatmore, 1997; Whatmore and Thorne, 1997). Ultimately, however ANT's symmetrical treatment of all storylines and actors poses difficulties in directly attending to the politics of ethical consumption. As Murdoch (1997, 367) has suggested, ANT 'merely concerns itself with which links hold and which fall apart. It thereby "flattens" all distinctions between the entities which comprise networks'. The approach is less helpful in understanding how networks might be 'negotiated by those who are only, at best, partial and ambivalent members' (Murdoch, 1997, 369). Insofar as the pursuit of ethical consumption strategies may require the disruption of relations between actors – not to mention moral judgments about actions – we need to be able to make distinctions between the strength and/or forms of power that particular actors may bring to and derive from networks.[2]

The politics of consumption in the home furnishings sector

Particularly in relation to other commodities, expressions of political concern about the composition, materials, processing and conditions of labour along the home furnishings commodity chain have been muted. To date, expressions of interest in the ethical consumption of furniture (e.g. through boycotts) are relatively underdeveloped.[3] There has been some interest in 'green' design, particularly from furniture designers seeking to work with non-traditional materials (Burall, 1991; Mackenzie, 1997; Press, 2001) and home furnishings magazines occasionally have highlighted 'eco issues' for consumers (Bailey, 1998; Silver, n.d.). Nonetheless, concerns which have prompted the development of new products, production methods and sourcing strategies in the food and paper products sectors (Barrett *et al.*, Chapter 1, this volume; Kishino *et al.*, 1999; Morris and Young, Chapter 4, this volume; Murdoch and Miele, 1999) have been less prominent in the context of furniture. Neither has attention been drawn to labour practices, as it has in the clothing sector (Crewe, Chapter 10, this volume). Perhaps the most obvious exception has been environmental marketing strategies pursued by the global furniture retailer Ikea, which we discuss in further detail below.

Home furnishings consumers we interviewed could be quite emphatic about their lack of concern for the geographical origins of products and the environmental impact of wood harvesting. A professional Canadian couple in their late 20s stated that they were unlikely to ask about environmental issues when shopping for furniture: '...not at all. Not that it wasn't important but we never asked.' An early thirties female consumer recalled her grandmother's shopping strategies:

> I remember even when we were young she would take us out to buy birthday presents...and she'd always look at the tag. 'Oh it was made in Taiwan. Nope. Find something made in Canada. And I hated that because it was so limiting' [...] I like to find out what types of woods are... I like to find out what it is. But as concerns the trees...in fact I would still buy teak [even knowing that it is endangered].

Although British and Canadian campaigns against the use of Southeast Asian teak in garden furniture were prominent in the mid-1990s, their longer term effect appears to have been relatively weak. In the United Kingdom, teak and mahogany have been substituted by sapele wood – which also is logged on a large-scale basis, often illegally.[4]

Consumers often suggested that the obligation to foreground the environmental provenance of furniture lay with retailers. A late twenties male Canadian consumer noted that retailers 'don't [...] really label furniture as environmentally friendly or environmentally unfriendly', while a female consumer pondered:

> What kind of paint did they use and how did they weld it and what kind of finish and what kind of wood and what kind of glue and who did it and how old were they and what kind of crate was it in? Who shipped it? Was it by boat or plane or train? [...] it's not like those people that are manufacturing it are going to put a label on it and tell you. And I think that's the only way. You know you can read a label off a can of food and see what is in it. You can decide whether you want to eat it. You don't get a label on a piece of furniture that lets you decide whether or not you want to purchase it for your home.
>
> (Mid-thirties, Canada)

The explicit contrast drawn with the food sector suggests consumers' growing familiarity with food labelling in many western countries.

Retailers reported that furniture consumers were unlikely to ask questions when shopping:

> [...] attitudes are changing but I don't think it's a groundswell. Nobody is coming in and saying they're boycotting you because you're selling wood furniture...It is not enough for me to go on a big green campaign...that said we only buy furniture that is environmentally friendly.
>
> (Buyer, Canadian department store)

An absence of explicit labelling or marketing strategies was thus explained by a (perceived) lack of interest by consumers in such practices. There was some suggestion from retail buyers that Canadian consumers might be more aware of environmental issues than UK consumers:

> the UK population is somewhere behind other parts of the world, be it Canada or somewhere else. We don't tend to be asked by our UK customers where things come from, how they're made, where they're produced. But I think that will be coming through more [...] We definitely will catch up with the Germans and the Swiss. They tend to be the ones who are keen on these issues.
>
> (UK mass market chain retailer)

Such comments might have been prompted – at least in part – by retailer interviewees' knowledge of our own biographies.[5] In Canada, the visibility of declining employment in the forestry sector alongside environmental protests against the clear-cut logging of 'old growth' forests (Willems-Braun, 1997) appears to have stimulated some consumer awareness of resource dependency within national and regional economies. Interviews typically did bear out the sentiments of a UK retail buyer who suggested that an absence of significant domestic wood resources had tended to reduce UK consumer interest either in domestically manufactured furniture or in the 'environmental' provenance of wood. Ultimately however it is difficult to advance strong generalizations about differing national consumption practices and for the purposes of this chapter we evaluate consumer concern across the two countries.

Although it is possible to identify a general lack of concern, some interviewees did express an interest in wood origins. Such consumers often suggested that they were uneasy about the manufacture of wooden furniture:

> I don't want to buy stuff that's made from rubber trees because it's coming out of Indonesia. Or Malaysia it's being you know cut down at an alarming rate so yeah…If I can I'd like Canadian.
>
> (Female, late thirties, Canada)

Concerns about wooden furniture are perhaps not surprising given a direct (and highly visible) connection with forestry and nature. For decades, environmental campaigning by organizations such as Friends of the Earth, Greenpeace and Global Witness has emphasized the problematic nature of forestry practices in the North and the South. In Canada public attention has been drawn to the negative effects of domestic clear-cut logging, while in Europe (where so-called 'ancient' forests were felled long ago) the loss of tropical rainforests has been higher on campaigners' agendas (see Vidal, 2002).

Although consumers give some thought to the geographical origins of different types of woods, assumptions about environmental 'friendliness' quickly become differentiated along North/South; northern/tropical; or softwood/hardwood lines. Occasionally interpretations are punctuated (particularly in Canada) by assumptions about the superiority of Swedish forestry practices relative to other northern countries; however negative stereotypes about wood production in 'distant lands' such as Indonesia and Malaysia persist. This othering of southern forestry itself presumes that intensively managed forests have environmental benefits. Yet Friends of the Earth (1997, n.p.) has argued that '"factory forests" are of far less value [than old growth] for wildlife or local inhabitants', and that plantation forestry in Scotland and Finland has been destructive to 'natural' peatlands. Further, although Canadian consumers might view domestic wood sources favourably, the British retail lumber chain B&Q boycotted Canadian softwood products in 2001 because they were not certified by the international certification organization, the Forest Stewardship Council (FSC) (CBC, 2001).[6]

The complexities of wood origins become even more involved when labelling and sourcing practices are scrutinized:

> right now what we are asking our suppliers, because in most of the cases it is very hard to find out, but we are asking them to provide us [with]... information. Where the heck is your lumber coming from? Because that is very tough to pinpoint and that is the first order of business towards any sustainable forestry issue... Do you know the region? Do you know [the] specific forest?
>
> (Mass market retailer, Canada)

Some manufacturing firms and smaller-scale designer-makers were highly sceptical of their ability to determine the geographical origins of the wood they used:

> I haven't really gone to big trouble in terms of getting the paper or getting the source. [...] I just specify that I want Canadian maple, so I'm hoping that's what they're shipping.
>
> (Small design-led manufacturer, Canada)

Although FSC and ISO 9000 certification systems attempt to foreground origins, the ultimate traceability of wood remains problematic in part because of the highly complex and contested processes bound up with particular forest management systems (see Corbridge and Kumar, 2002). Certification thus provides only a very partial means of regulating timber supply. Further, certification 'introduces policy changes through commercial rather than central [government] or local power and uses market acceptance rather than regulatory compliance as an enforcement mechanism' (Naka *et al.*, 2000, 476). Retailers often are able to quickly capitalize on certification systems, as demonstrated by B&Q's profession of leadership in establishing the FSC (B&Q, 1995, 2002). If manufacturers and designer-makers are uncertain about wood biographies, consumers with fewer available insights into material origins are presented with further difficulties. Differential power between actors in commodity networks ultimately may undermine attempts at traceability.

Some consumers indicated a considerable knowledge of the economic geographies of British and Canadian manufacturing. In the context of a long term decline in domestic furniture industries and expanding import penetration (Industry Canada, 1996; DTI, 1995), both UK and Canadian consumers problematized the purchase of imported goods:

> especially the wood, it bothers me that they ship it out of Canada as raw export, and then... then we lose all the jobs. They can make something here.
>
> (Female consumer, mid-forties)

Here there is a particular reference to the resource-dependent nature of the Canadian economy and a desire for value-added production. In the

United Kingdom, Beth was particularly proud of pieces she and her husband had purchased at a local retailer. A Suffolk-based manufacturer had provided retailers with an information label highlighting the 85-year history of the firm and its sourcing strategy 'from our own wood supply'. Beth (early fifties) kept the label in the top drawer of her chest in the living room:

> it's nice to know that you've got English things. I know that's silly in one way, but it's nice to know that we're keeping the English traditional furniture going. Because if a lot of people all said 'oh well we're going to Ikea' or whatever, there'd be no joiners or handcrafters or things in England, would there? [...] I know you pay more for it, but we just feel that we're just giving a bit back and keeping it going.

The sensibilities of consumers such as Beth stand in contrast to perceptions held by retailers. In this example, manufacturer, retailer and consumer are bound up in a web of connections which underscore skilled domestic production alongside knowledgeability about wood origins.

Ethical consumption practices

In bracketing together concerns about labour practices, product composition and the environmental consequences of production and distribution, we have sought to emphasize the multitude of ways in which consumers frame ethical consumption (Bedford, 2000; Eden, 1993; Miller, 1995). Faced with complicated and potentially conflicting options, some consumers choose to detach themselves from these concerns in favour of a focus upon seemingly more straightforward issues such as price. Following a suggestion that retailers did not present a choice between wooden furniture from 'managed' versus 'regular' forests, a late twenties Canadian couple noted that 'I think the thing is that also we aren't willing to pay extra money either'. Comments such as these suggest that while consumers may be aware of furniture production processes and geographies, they still tend to express consumption decisions as monetary choices.

Confronted with varied concerns about specific materials; origins of components; and the problematic nature of shipping goods long distances, it may be that consumers are seeking to simplify decision-making processes:

> I think it's so complicated that it's almost impossible to make the right choice. [...] I certainly don't feel that I know enough to be able to claim that you know that a plastic chair is better or worse than this wooden chair. [...] You know I can suspect that if it was made locally it's probably better because, you know, there's less transportation you have to look at, where all the materials are coming from.
>
> (Male consumer, forties, Canada)

Another consumer suggests that it is difficult to make decisions about products which go through multiple stages of production:

> I find it really hard because there are so many different phases to the manufacturing of this table that it becomes daunting as a consumer to be socially conscious about everything and you just want to fit it in your living room... So you don't want to think about it as much. I think with food, you can be a little more socially conscious because it is something that you are consuming internally and there is not as much processing factors that a head of broccoli goes through.
>
> (Female, mid-thirties, Canada)

In contrasting tables with broccoli, the consumer draws attention to the importance of distinct spatialities associated with different commodities.

Furniture and the body

Labelling and marketing practices by food retailers particularly of branded 'green' or 'organic' products (Guthman, Chapter 12, this volume; Morgan and Murdoch, 2002) – are both influenced by and in turn act to influence consumer concerns about environmental sourcing. A contemporary sense of rupture in ontological security towards food purchases is closely connected to the direct ingested relationship between food and the body (Whatmore, 1997). Furniture manufacturing utilizes a range of 'natural' materials such as wood and metal but also often incorporates an array of synthetic materials and chemicals. Yet consumers were relatively unreflective about the presence of volatile organic compounds (VOC) in solvents and coatings and toxic substances in particleboard. As a Canadian designer noted:

> in furniture there are a lot of materials that are synthetic materials that offgas... Even though they look like they're solid and they're stable. They are not, and a lot of these chemical vapours that are off-gassing are harmful. So in a way it is not different than ingesting a strawberry with pesticides on it. People that are inhaling fumes all day long are going to have less of tolerance over these kinds of things and they're going to be a lot more vigilant in terms of the kinds of materials that are used in furniture. And I think those companies that start to anticipate those kinds of things and solve those issues and promote those solutions are going to be received by consumers in a much more favourable manner than those who ignore it and deny that it even exists.

The existing perception of a more detached relationship between furniture and the body disrupts associations between furniture consumption and environmental hazards. As yet there appears to have been little impetus for furniture manufacturers to shift to environmentally 'friendly' production processes or materials. The technological development of water-soluble finishes, for example, has proceeded extremely slowly. This disconnection of production from consumption is not surprising given the relative distantiation of the product from the body, and also is apparent in sectors such as the auto industry. Pressures to develop water-based automotive paints and lacquers have stemmed more from national or international requirements to reduce environmentally harmful waste water outputs and atmospheric emissions from large-scale production sites, rather than a direct concern for the bodies of final consumers.

Despite the fact that certain materials may pose health concerns, the perception of personal risk by consumers is minimal. A male consumer in his 30s whose home was furnished with second-hand plastic and fibreboard pieces suggested that any 'environmental' aspects of his furniture re-use were insignificant – he was more interested in the 1950s styling of objects than the 'recycling' of products.[7] He was relatively unconcerned about the materials used in construction:

> if I was like that way I wouldn't be buying [that kind of stuff]…because actually a lot of the stuff that this stuff is made of is pretty hazardous…

Negative reactions to 'synthetic' or 'artificial' materials in furniture or finishes generally tended to be expressed in terms of the quality and value of solid wood, rather than bodily risk:

> I like the sense that something that is solid wood as opposed to MDF [medium density fibreboard] or something like that. It feels cleaner and more solid…and has more value. But I don't think it is in my mind… I'm not driven by those environmental concerns, although I suppose, if I thought about it I might. It just has never really crossed my mind.
>
> (Female, forties, Canada)

In contrast, notions of quality in food are much less distantiated from the body. Although they may be affiliated with localized food networks – which are often positioned as 'alternative' to 'industrialized' production (Murdoch *et al.*, 2000) – 'quality' issues in food consumption often are most immediately associated with ingestion and health. A direct connection to bodily well-being may more easily politicize issues surrounding food origins and composition. Even within the food sector itself there is evidence that consumer concerns about food safety are greater than worries about broader environmental impacts of pesticides use. Raynolds (2000, 304) suggests that organic trade in fresh fruits and vegetables – particularly bananas – stems primarily from 'rising health and safety concerns and to a lesser extent […from] environmental concerns among Northern consumers'. Thus consumption practices appear to be primarily rooted in concerns about a direct impact upon the consumer body through ingestion.

Interestingly too, although notions of food quality may be associated with an absence of negative characteristics such as genetic modification, the use of additives in processing or hormones in meat production, food 'quality' also derives from a range of positive attributes such as nutritional values, freshness and cleanliness, which are seen to be beneficial to the body. This directly positive connection between consumption and bodily health has few parallels in the furniture industry, apart from posture-enhancing chairs or mattresses which use 'technologically advanced' materials to encourage comfortable sleeping.

Durability, quality and meaning

Differing attitudes towards environmental sourcing between furniture and other products such as food or clothing also derive from furniture's relative durability. Actors across the chain suggest that the longevity of furniture leads to a reduced concern for its environmental biography:

> I think that the sort of level we are working at, even if we were to be selling solid mahogany I wouldn't have a problem with it because I would know that the piece of furniture is going to last 100 years or more. It is a very good use of that resource.
>
> (UK high-end independent furniture retailer)

Similarly, a Canadian furniture designer suggested that:

> the way I see my products is that they're not like plastic products where you throw them out after [...] You can hand them down to the next generation so they are really credible products and, you know, you can refinish it if you scratch the wood...[...] And in that way I think they're much more ecological, even though they use wood.

Environmental concerns are thus mediated by the physical properties of objects:

> [Furniture is] not disposable. You know, I'm going to buy a chair and yes, an animal may have died to make that leather chair, and there's wood in it, but I'm going to use that chair for years...hopefully and so I don't think of it as being wasteful. And the thing about food packaging or paper products is the wastefulness of it.
>
> (Female consumer, early thirties, Canada)

Although consumers may make connections between disposable paper products and unsatisfactory forestry practices, the distinctive temporality of furniture reduces an immediate concern about origins and production.

As Shove and Warde (2002, 245) have argued, 'the social and physical durability of objects is of real significance. It matters little how "green" that settee is if it is discarded and replaced after only two years'. In locating the 'green' consumer, they review a series of mechanisms which drive expanding levels of consumption. These include social comparison, the creation of self-identity, mental stimulation (novelty), the 'matching' effect and specialization within daily life (Shove and Warde, 2002). Most importantly, Shove and Warde (2002) argue that each mechanism will often take on a dual role. That is, we must consider not only consumers' selection between different products at the point of consumption, but also we should explicitly examine the 'churn' rate of products: 'the rate at which things are replaced, demolished and thrown away' (Shove and Warde, 2002, 245). While the narratives we outline here may not always emphasize an initial selection of 'green' products, consumers have foregrounded the importance of the longevity of furniture and a consideration about the rate at which products are replaced.

> we bought that [wardrobe] to go into a brand new bungalow and at that time Ercol was really big, still is big, I mean it is solid furniture it is solid wood. And it is good quality [...] And there are no veneers to peel off [...] We just liked the feel of it and the fact that it was solid wood.
>
> (Female consumer, late fifties)

Consumers often contrast solid wood to composite constructions such as veneers. Here, associations are made between solid wood, a branded product and the resultant durability of the piece.

It would be too simplistic to suggest that a lack of concern for environmental origins represented an absence of interest in the broader provenance of goods: product biographies (Cook and Crang, 1996) often are very important to consumers. Vince and Donna's (late fifties) most recent and valued purchase, for example, was a nineteenth century wooden cabinet originally manufactured in Quebec:

> there was this certain appeal to it just because it was old. You could maybe feel the energy out of it, I don't know. When you walked into this guy's shop and saw the extent of all these at one time, grand pieces of furniture and they'd been lovingly stored and been prepared for re-use. It was just a wonderful feeling. A good experience. Not [much] furniture being built today will have that reverence. Maybe something [like our other sideboard] which is solid wood like that will be around for a long time, but a lot of this stuff that's pressed board and veneer isn't worth...

Vince emphasizes that the longevity of the cabinet has facilitated its 'recycling' for a new user, and contrasts it positively with more disposable furniture made of less resilient materials.

Environmental consumption and the role of the retailer

Furniture has a particular relationship with human bodies in space and across time. The absence of ingestion – despite bodily contact – in this relationship appears to have blunted consumer concern with the composition of final products. At the same time, the longevity and durability of furniture has tended to mute an overt questioning of raw material origins or labour conditions in furniture manufacture. Where consumer anxieties do surface they generally are related to the use of tropical hardwoods and issues surrounding 'local' versus 'foreign' production. Although some consumers express a certain knowledge of production processes, many ultimately fall back upon considerations made at the immediate point of purchase, such as price.

Furniture retailers often insist that there is insufficient consumer awareness to warrant attention to environmental issues. Where retailers do take note of consumer concern about environmentally 'friendly' or ethical policies, they relate their actions to future or potential expectations, rather than to a current vocal demand by consumers:

> I think if you are a large company and you have a reputation to uphold with the public, they have to be confident that what they are buying is not coming from a country with child labour so that they can trust what they buy and I think particularly in home products it is in some ways even more important because they are products that are going to be used to cook their food in, eat off and they have to be confident that nothing is going to happen to them when they do.
>
> (UK department store buyer)

The buyer not only conflates different types of ethical and environmental concerns, but also emphasizes that concern is likely to be greater where commodities have close contact with the body.

Our final section reflects upon a case in which the politics of ethical retailing and consumption are highly visible. It is no coincidence that the example is one of a self-defined 'lifestyle' retailer. Lifestyle retailers may be more likely to respond to consumer demands, as they seek directly to appeal to values held by their clientele and to sell products, images and ethics congruent with those values. One buyer explicitly considered her company to be:

> a lifestyle retailer rather than just a product retailer. So we are very interested in … what is our customer actually doing at the moment and what is influencing the decisions that they are making? [...] And why is it suddenly that everyone is becoming [...] environmentally kind of aware? It is just those kinds of things that impact on people's environment that actually influence what they believe and then we fulfil that need.
>
> (UK furniture retailer)

The pursuit of ethical marketing strategies by lifestyle retailers must be seen in the light of broader competitive tactics. This can be seen most clearly in the case of the global furniture retailer Ikea, whose 'green' production, marketing and advertising strategies can be more appropriately explained as an attempt to achieve dominance within the retail hierarchy rather than as a direct response to consumer demands.

Founded in 1943 by Ingvar Kamprad in a small town in southern Sweden, Ikea now has 143 stores in twenty-two countries (Ikea, 2002). While Sweden, Germany and Italy remain important sources of supply, Ikea increasingly sources

Table 13.1 Ikea's global reach

Share of global sales, 2002 in %		Share of global production, 2002 in %	
Germany	21	Sweden	14
United States	13	China	14
United Kingdom	12	Poland	8
France	9	Germany	6
Sweden	7	Italy	6

Source: Adapted from Ikea, cited 2002.

products from lower cost countries. Purchasing from China, for example, increased from a 10 per cent share in 2000 to 14 per cent in 2002 (see Table 13.1).

National web sites and catalogues seek to extend knowledge to consumers about wood sourcing policies, supplier requirements and materials used in furniture and household goods. The company deploys a performance of environmental sustainability and responsibility in which consumers are able to obtain inexpensive products 'but not at any price' (Ikea catalogue, 2002). Emphasis also is placed upon the retailer's role in developing innovative designs as a means of finding 'new ways to keep prices low' (ibid.). Multiple dimensions of ethical trading are fused in the pursuit of a 'triple bottom line', in which 'both environmental and social responsibility [are incorporated] into business objectives' (Hughes, 2001, 396).

Particularly in the context of wooden products, Ikea seeks to play an active role in alleviating knowledge deficit among suppliers and in supporting 'sustainable' forestry by providing information to consumers. The retailer supports Global Forest Watch, a mapping initiative that identifies the location of intact forests, their ownership and the companies logging these forests (Sizer, 2000). Such maps enable Ikea (in conjunction with its suppliers) to source wood from responsibly managed forests (Sizer, 2000, 75). There are clear parallels between such attempts to trace wood 'back' to its forest of origin and quality assurance schemes in the food industry which claim to trace meat to individual farms (Morris and Young, Chapter 4, this volume). The aim is directly to reconnect consumers and producers.

However as we have noted above, recovering wood origins is not a straightforward task. By 2005, Global Forest Watch aims to publish data for twenty-one countries (covering three quarters of all 'frontier forests'), but information currently is available for only eight countries (GFW, 2002). The Forest Stewardship Council certification scheme provides a further marker, although Ikea's current minimum requirement is that suppliers have FSC certification 'or an equivalent system' (Ikea, 2002) thus diluting even the regulatory force of FSC guidelines. Ikea (2002) acknowledges that:

a large proportion of Ikea products come from suppliers in countries where environmental work is, on the whole, well developed, but we also purchase products from countries where environmental work has not yet proceeded quite as far.

Their declared priority is to work with suppliers in countries that do not have well developed environmental legislation in order to improve local conditions. Such an approach – working *towards* rather than necessarily achieving environmental responsibility – could function to absolve Ikea of blame if a supplier was found to have poor working conditions or unsatisfactory environmental management systems.

Ikea also emphasizes its leading role in reducing environmental and bodily hazards associated with furniture production and consumption. The firm has been involved in Greenpeace-collaborative projects to eliminate chlorine bleach from paper used in catalogues and to ban the hardening agent polyvinylchloride (PVC) from plastic products. The global adoption of German environmental regulations banning formaldehyde and pentachlorophenol (PCP) reflects not only the importance of that country to Ikea's sales (see Table 13.1), but also the extent to which the firm is sensitive to geographic variations in knowledges about perceived hazards. In the United Kingdom, Ikea's compulsion that suppliers manufacture low-formaldehyde chipboard resulted in an industry-wide transformation in product specification. Ikea thus has sought explicitly to negotiate geographical differences in knowledges about environmental concern (whether expressed by individual consumers or through national regulatory practices). At the same time, Ikea has been able to position itself at the forefront of retailer responses to environmental issues.

Ikea also has sought to respond to negative publicity about the use of child labour (*Canadian Press*, 1997; *Macleans*, 1995). In 1998 a highly publicized agreement was signed with the International Federation of Building and Woodworkers, regulating working conditions for employees involved in the manufacture of wooden products for Ikea (IFBWW, 2001).[8] The agreement now covers both independent suppliers and factories owned by Ikea and demands compliance with national labour legislation and ILO conventions. Workers must have rights to free collective bargaining; child labour, forced labour and nonvoluntary work in prisons are prohibited, as are gender, racial or political discrimination. Adherence to health and safety and environmental legislation also is covered in the agreement. Standards are monitored by groups which include two Ikea and two IFBWW members. Recent visits to factories in eastern Europe and southeast Asia revealed generally good working conditions and acceptable environmental standards, particularly in factories where Ikea was a bigger share of the manufacturer's total production (IFBWW, 2001).

Ikea's code of ethics is unprecedented in the furniture sector, but parallels codes developed within clothing and fashion retailing (Hale and Shaw, 2001; Johns and Vural, 2000). Whether or not corporate codes facilitate improved working conditions remains debatable (see in particular Hale and Shaw, 2001). In addition to practical difficulties involved in monitoring (Hughes, 2001), many activists view the imposition upon Southern nations of ethical codes developed in the North as problematic (Elmhirst, pers. comm.; Hale and Shaw, 2001) – and workers themselves often have little input. The incorporation of international trade unions such as the IFBWW into networks of actors represents a positive development, although recent discussions of the potentials of global unionism have not been unreservedly optimistic about the current position of organized labour as set against multinational capital (Waterman and Wills, 2001).

Like many clothing and footwear manufacturers, Ikea regularly switches suppliers in order to obtain the lowest cost. This raises concern about its ability to influence working conditions in the medium- or long-term. A Canadian designer involved in promoting domestic furniture manufacturing in Trinidad reported that:

> You go into Ikea and you see some pretty cheap stuff.... You work backwards.... So [Ikea's code of conduct] could easily be more of a marketing position than a reality.... Ikea had been in Trinidad before I got there and just killed a couple of companies [when they left].

Consumers often are aware that low prices can signal problematic labour relations at the point of production. Below, Denise (late thirties) implies that terms and conditions in Canada are better than other countries:

> I don't know about some furniture. The big furniture companies – like you know, [inexpensive mass market retailer] or whatever – to know how much of their products are Canadian. Probably not very many of them if they're offering cheap prices.

There is a heightened scepticism towards the claims of retailers with respect to ethical sourcing, particularly when consideration is given to retail price levels.

As a truly global retailer, Ikea's potential influence upon patterns of international production and consumption is significant. On the one hand, the continuing development of furniture as a 'lifestyle' commodity may well increase levels of environmental concern among consumers, compelling other retailers to alter sourcing strategies. In September 2001 the US retailer Pottery Barn announced its withdrawal from Burma, largely in response to the Free Burma Coalition's threat to hold protests at outlets in six US cities (*Far Eastern Economic Review*, 2001). Ikea's global reach has facilitated political possibilities across the chain: it may be easier for unions to challenge practices along commodity chains over which retailers exert significant influence.

However our discussion has emphasized two important aspects of the Ikea case. The first is designers', manufacturers' and consumers' scepticism about the traceability of product biographies. Given significant difficulties in monitoring and auditing global production chains, it is perhaps not surprising to find expressions of disinterest from individual consumers about the ethical purchase of furniture. The second issue relates to the ways in which power relations have worked themselves out across global production networks. Although 'notions of business responsibility [may be] transforming the organisation of production' (Hughes, 2001b, 401), such transformations are proceeding through – and precisely *because of* – retailer-dominated supply chains. Such strengthening of retailer power has important political implications.

Conclusions

The chapter traces flows of information and knowledge across the home furnishings commodity network. In order to understand connectivities between actors, we have foregrounded consumer knowledges about labour practices, the origins of materials, the construction of products and the resulting geographies of commodity chains. At a broad level our account of the 'ethical' consumption of furniture parallels other studies which have stressed 'the complexities and ambiguities in the relationship between environmental knowledge and behaviour' (Eden, 1998, 430). Further, we have attributed relatively low degrees of consideration and action to the distinctive spatialities and temporalities of furniture, namely its relationship with the body and its relative durability.

Concern with corporate image has led 'lifestyle' retailers such as Ikea to maintain an environmental profile as part of public relations campaigns. Given the potential 'disposability' of fashion- and design-led products (Leslie and Reimer, 2003), it may become more important for retailers to assuage fears relating to the conditions of production and hazards associated with furniture and other household products, because such goods have come to be perceived as less enduring. For consumers, awareness of the environmental consequences of consumption is likely to increase. The spatial connectivity between furniture and bodily health may become more visible as we learn more about the hazards and consequences of furniture consumption on the landscape and on our bodies.

However, our examination of the politics of consumption in the home furnishings commodity chain points to potential difficulties with recent exhortations that increased consumer knowledge provides the basis from which the commodity chain can be transformed. Hartwick (2000, 1178), for example, has argued that 'by focussing on human actors at points along networks specified as commodity chains we can find a new basis for forging innovative radical political tactics'. It may well be that home furnishings consumers face a 'lack of agency and empowerment in the face of environmental concern' (Eden, 1998, 430). Yet consumption practices emerge from a complex set of interactions and negotiations by a variety of agents located at different nodes. As a result, the 'demystification' of origins, or the pursuit of geographical 'detection' (Hartwick, 2000, 1184) cannot be read as a straightforward task. Interrogating cashiers about product origins, for example (Hartwick, 2000, 1189), side-steps the position of retail workers themselves – of their wages, terms and conditions of employment, and their ability to influence flows of knowledge, information and power.

The more globally extensive a network, the more likely it is that producers and retailers will be compelled to develop ethical codes of conduct. Given the potential for geographical variation in consumer awareness, the retailer operating in the most countries will draw upon more diverse knowledges and will be more likely to adopt ethical policies. At the same time, such conditions can enhance rather than challenge existing power relationships between retailers, producers, workers and consumers. As the Ikea example demonstrates, ethical connectivities between sites might be strongly orchestrated by retail actors for competitive purposes. This is not to say that power relationships are always already stabilized – rather, they are the product of a laboured, uncertain and contested process (Whatmore and Thorne, 1997, 290). Precisely because interdependencies between different actors are unstable and dynamic, applying political pressure to one site will not necessarily cause signals to pass easily 'up' or 'down' the chain. Forging an ethical politics along the furniture network necessitates the tracing of complex linkages, flows and interdependencies between different actors as well as a recognition that opposition may be feasible only at certain moments and locations.

Acknowledgements

Research on the home furnishings commodity chain was funded by the Economic and Social Research Council (Award No. R000237580) and the Social Sciences and Humanities Research Council of Canada (Award No. 410-97-0335). Many thanks to Sally Eden and Alex Hughes for insightful comments on a previous draft.

Notes

1 The research upon which this chapter draws included 160 interviews with furniture manufacturers, retailers, designers, magazine editors and consumers in the United Kingdom and Canada. We were particularly interested in the dynamics of innovative, design-led retailers and manufacturers. Having traced consumers through networks which originated with such retailers, interviewees were predominantly middle to upper-middle class urban dwellers who are by no means (nor did we intend them to be) 'representative' of all home furnishings consumers. Names have been changed to protect the anonymity of participants.
2 We are indebted to Sally Eden for pushing us to clarify our thoughts on ANT in relation to ethical consumption.
3 This is not to overlook specific campaigns relating to wood and lumber. Since the late 1980s Friends of the Earth has published information on timber supplies (particularly from tropical forests) and reclaimed wood in *The good wood guide*. *The green consumer guide* (Elkington and Hailes, 1988) also included a section on 'choosing acceptable wood products'.
4 In 2000, up to 60 per cent of all UK tropical timber imports were thought to be illegal (Vidal, 2002).
5 Part of our original interest in the furniture industry can be traced to postgraduate student days at the University of British Columbia, at a time when anti-clearcut logging protests were gathering strength and when the absence of value-added production was becoming increasingly evident.
6 The aim of the Forest Stewardship Council (FSC) is to set a common standard for forest management by promoting a recognized set of principles and a consistent mechanism for evaluating forestry practices (Simula, 1997). Other forms of timber certification include the international standard ISO 14000, which certifies suppliers' 'management structure and environmental conditions on a facility-by-facility basis' (Naka, 2000, 475) and ISO 9000, which is more specifically concerned with 'environmental management' (Crossley *et al.*, 1997; Handfield *et al.*, 1997).
7 There is insufficient space here to explore possible differences in 'brand new' versus second-hand and antique markets for furniture. Although the main focus of our original research was upon new products, the consumption end of the chain is

rendered more complex by the multitude of ways in which furniture may be acquired for the home.

8 IFBWW has a membership of over 11 million workers organized into 281 trade unions in 124 countries (IFBWW, 2001).

References

B&Q (1995) *How Green is My Front Door? B&Q's Second Environmental Review.* Eastleigh: B&Q.

B&Q (2002) B&Q and the environment. Leaflet available at www.diy.co.uk, accessed 1 August 2002.

Bailey, S. (1998) The green house effect. *Elle Decoration*, September, pp. 50–4.

Bedford, T. (2000) Ethical consumerism: everyday negotiations in the construction of an ethical self. Unpublished PhD thesis, Department of Geography, University College London.

Burgess, A. (2001) Flattering consumption: creating a Europe of the consumer. *Journal of Consumer Culture*, 1, pp. 93–117.

Burall, P. (1991) *Green Design.* London: Design Council.

Canadian Press (1997) Canadians tackle child labour without sanctions while U.S. proposes ban on imports. *Canadian Press*, 24 October, n.p.

Canadian Broadcasting Corporation (2001) *Marketplace: 'Good wood'* Programme originally aired 2 January 2001. Transcript available at www.cbc.ca/consumers/market/files/environ/goodwood/index.html, accessed 23 May 2002.

Cook, I. and Crang, P. (1996) The world on a plate: culinary culture, displacement, and geographical knowledges. *Journal of Material Culture*, 1, pp. 131–53.

Corbridge, S. and Kumar, S. (2002) Community, corruption, landscape: tales from the tree trade. *Political Geography*, 21, pp. 765–88.

Crang, P. (1996) Displacement, consumption and identity. *Environment and Planning A*, pp. 47–67.

Crang, P. (2000) Organisational geographies: surveillance, display and the spaces of power in business organisations. In Sharp, J., Routledge, P., Philo, C. and Paddison, R. (eds), *Entanglements of Power: Geographies of Domination/Resistance.* London: Routledge, pp. 204–18.

Crossley, R., Primo-Braga, C. and Varangis, P. (1997) Is there a commercial case for tropical timer certification? In Zarilli, S., Jha, V. and Vossenaar, R. (eds), *Eco-labelling and International Trade*. London: Macmillan, pp. 228–50.

Department of Trade and Industry (1995) *Furniture Manufacture: Growth Through Excellence*. London: DTI.

Eden, S. (1993) Individual environmental responsibility and its role in public environmentalism. *Environment and Planning A*, 25, pp. 1743–58.

Eden, S. (1998) Environmental issues: knowledge, uncertainty and the environment. *Progress in Human Geography*, 22, pp. 425–32.

Elkington, J. and Hailes, J. (1988) *The Green Consumer Guide*. London: Victor Gollancz.

Far Eastern Economic Review (2001) Pottery Barn bins Burmese goods. 11 October, p. 10.

Fine, B. and Leopold, E. (1993) *The World of Consumption*. London: Routledge.

Friends of the Earth (1997) Wood and environmental space: briefing http://www.foe.co.uk/pubsinfo/briefings/html/19971215145244.html, accessed 14 December 2001.

Global Forest Watch (2002) About Global Forest Watch http://www.globalforest-watch.org/english/about/index.htm, accessed 1 August 2002.

Goodman, D. (1999) Agro-food studies in the 'age of ecology': nature, corporeality, bio-politics. *Sociolgia Ruralis*, 39, pp. 17–38.

Goodman, D. (2000) Organic and conventional agriculture: materialising discourse and agro-ecological managerialism. *Agriculture and Human Values*, 17, pp. 215–19.

Hale, A. and Shaw, L. (2001) Women workers and the promise of ethical trade in the globalised garment industry. *Antipode*, 33, pp. 510–30.

Handfield, R., Walton, S., Seegers, L. and Melnyk, S. (1997) 'Green' value chain practices in the furniture industry. *Journal of Operations Management*, 15, pp. 293–315.

Hartwick, E. (1998) Geographies of consumption: a commodity chain approach. *Environment and Planning D: Society and Space*, 16, pp. 423–37.

Hartwick, E. (2000) Towards a geographical politics of consumption. *Environment and Planning A*, 32, pp. 1177–92.

Hughes, A. (2000) Retailers, knowledges and changing commodity networks: the case of the cut flower trade. *Geoforum*, 31, pp. 175–90.

Hughes, A. (2001) Global commodity networks, ethical trade and governmentality: organising business responsibility in the Kenyan cut flower industry. *Transactions of the Institute of British Geographers*, 26, pp. 390–406.

Ikea (2002) www.ikea.co.uk, accessed 9 August 2002.

Industry Canada (1996) *Household Furniture. Part 1. Overview and Prospects.* Ottawa: Industry Canada.

International Federation of Building and Woodworkers (2001) http://www.ifbwww.org, accessed 1 June 2001.

Jackson, P. (1999) Commodity cultures: the traffic in things. *Transactions of the Institute of British Geographers*, 24, pp. 95–108.

Jackson, P. (2002) Commercial cultures: transcending the cultural and the economic. *Progress in Human Geography*, 26, pp. 3–18.

Johns, R. and Vural, L. (2000) Class, geography and the consumerist turn: UNITE and the Stop Sweatshops Campaign. *Environment and Planning A*, 32, pp. 1193–213.

Kishino, H., Hanyu, K., Yamashita, M. and Hayashi, C. (1999) Recycling and consumption in Germany and Japan: a case of toilet paper. *Resources, Conservation and Recycling*, 26, pp. 189–215.

Klein, N. (2000) *No Logo.* London: Flamingo.

Leslie, D. and Reimer, S. (1999) Spatialising commodity chains. *Progress in Human Geography*, 23, pp. 401–20.

Leslie, D. and Reimer, S. (2003) Fashioning furniture: restructuring the furniture commodity chain. *Area*, 35, pp. 427–37.

Macleans (1995) Saving the brand name. 12 November, 108(50), p. 30.

Mackenzie, D. (1997) *Green Design*, Second edition. London: Lawrence King.

Miller, D. (1995) Consumption as the vanguard of history. In Miller, D. (ed.), *Acknowledging Consumption: A Review of New Studies.* London: Routledge, pp. 1–57.

Morgan, K. and Murdoch, J. (2000) Organic vs. conventional agriculture: knowledge, power and innovation in the food chain. *Geoforum*, 31, pp. 159–73.

Murdoch, J. (1997) The spaces of actor-network theory. *Geoforum*, 29, pp. 357–74.

Murdoch, J. and Miele, M. (1999) 'Back to nature': changing 'worlds of production' in the food sector. *Sociologia Ruralis*, 39, pp. 465–83.

Murdoch, J., Marsden, T. and Banks, J. (2000) Quality, nature and embeddedness: some theoretical considerations in the context of the food sector. *Economic Geography*, 76, pp. 107–25.

Naka, K., Hamnett, A. and Stuart, W. (2000) Forest certification: stakeholders, constraints and effects. *Local Environment*, 5, pp. 475–81.

Press, M. (2001) Crafting a sustainable future from today's waste. *The Interdisciplinary Journal of Design and Contextual Studies*, available at www.co-design.co.uk/mpress.htm, accessed 14 December 2001.

Raynolds, L. (2000) Re-embedding global agriculture: the international organic and fair trade movements. *Agriculture and Human Values*, 17, pp. 297–309.

Shove, E. and Warde, A. (2002) Inconspicuous consumption: the sociology of consumption and the environment. In Dunlap, R., Buttel, F. H., Dickens, P. and Gijswijt, A. (eds), *Sociological Theory and the Environment*. Lanham, MD: Rowman and Littlefield, pp. 230–51.

Silver, R. (n.d.) *Eco-stylists*. Originally published in *idFX*; available at www.trannon.com/idFX.htm, accessed 14 January 2002.

Simula, M. (1997) Timber certification initiatives and their implications for developing countries. In Zarrilli, S., Jha, V. and Vossenaar, R. (eds), *Eco-labelling and International Trade*. London: Macmillan, pp. 207–27.

Sizer, N. (2000) Forest futures. *Space*, 4, pp. 72–9.

Smith, N. (1990) *Morality and the Market: Consumer Pressure for Corporate Accountability*. London: Routledge.

Vidal, J. (2002) UK plays key role in illegal logging. *The Guardian*, 19 April.

Waterman, P. and Wills, J. (2001) Space, place and the new labour internationalisms: beyond the fragments? *Antipode*, 33, pp. 305–11.

Whatmore, S. (1997) Dissecting the autonomous self: hybrid cartographies for a relational ethics. *Environment and Planning D: Society and Space*, 15, pp. 37–53.

Whatmore, S. and Thorne, L. (1997) Nourishing networks: alternative geographies of food. In Goodman, D. and Watts, M. (eds), *Globalizing Food: Agrarian Questions and Global Restructuring*. London: Routledge, pp. 287–304.

Willems-Braun, B. (1997) Buried epistemologies: the politics of nature in (post)colonial British Columbia. *Annals of the Association of American Geographers*, 87, pp. 3–31.

Index

Printed and bound by CPI Group (UK) Ltd, Croydon, CR0 4YY

01/11/2024

01782615-0012